Carl McDaniel is a visiting professor at Oberlin College in Oberlin, Ohio, and a professor emeritus at Rensselaer Polytechnic Institute in Troy, New York. He was the founding director of the undergraduate environmental science degree program at Rensselaer. For the first three decades of his academic career, he researched insect and plant development. His scholarly interests since have been focused on the interface between biology and economics. He has written four books: *Paradise for Sale: A Parable of Nature* (2000, with economist John M. Gowdy); *Wisdom for a Livable Planet: The Visionary Work of Terri Swearingen, Dave Foreman, Wes Jackson, Helena Norberg-Hodge, Werner Fornos, Herman Daly, Stephen Schneider, and David Orr* (2005); *Trail Magic: Creating a Positive Energy Home* (2012) and *At the Mercy of Nature: Shackleton's Endurance Saga Gives Promise for Our Future* (2014, with illustrations by Donald Watson).

Spencer Berry, Anton Lang and Ian Sussex

Carl McDaniel

Beauty Won Me Over

A Scientist's Life

Austin Macauley Publishers™
London • Cambridge • New York • Sharjah

A CIP catalogue record for this title is available from the British Library.

ISBN 9781528986021 (Paperback)
ISBN 9781528986038 (Hardback)
ISBN 9781528986045 (ePub e-book)

www.austinmacauley.com

First Published (2021)
Austin Macauley Publishers Ltd
25 Canada Square
Canary Wharf
London
E14 5LQ

To begin with, I thank those whose names are mentioned in the text for all they did to make this book possible. I am indebted to Oberlin College's Environmental Studies Department for giving me an academic home and support for writing *Beauty Won Me Over: A Scientist's Life*. David Benzing, Norm Craig, John Elder, Bob Longsworth, David Sonner, and Marci Peterson kindly read early drafts, making innumerable suggestions that improved the manuscript immensely, and Don Watson critiqued the penultimate draft and wrote the Foreword. David Orr also provided invaluable commentary and found the title, *Beauty Won Me Over*, in the passage that described how I chose to work on an elegant moth, *Hyalophora cecropia*, for my master's thesis. I thank Linda Thomas for creating the exquisite image of the life cycle of *H. cecropia* that appears on the front cover. My thanks to Dale Preston for using his electronic skill in making the illustrations from 35 mm pictures I had taken, and thanks to the Michigan State Plant Research Laboratory in East Lansing, Michigan, for the picture of Anton Lang.

I cannot say enough about Kelly Viancourt, who thoroughly edited the best draft I could produce. With skill, she cut the unnecessary and used her superb editing ability to make what remained into a coherent, readable text.

Mary, my wife, witnessed and participated in the gestation and birthing of five books from her husband, who never intended to write even one. Our friends call her a saint and she is.

The people mentioned above, and others, did their best to have me write only that which is true, but errors and misstatements remain for which I am responsible.

Table of Contents

List of Illustrations

Foreword

Beauty Won Me Over: A Scientist's Life provides a record of a life lived. Its stories—along with incidents and accidents of serendipity on the life path of a determined scientist—offer what any reflection of the past can best give: lessons for the future. The details, with charming memories that any of us may have had but have long forgot, are made accurate in McDaniel's biography by records kept, evidently in boxes of papers, a model habit for anyone, especially future writers, "to keep notes."

The author thus tells us of the milestones of his career, but also side stories: a "wise guy's" firecracker that caused a false start of the Boston Marathon in 1965. Or, as a newly groomed naval officer on the destroyer *USS Anderson*, how he would enlighten the crew with dissertations on the biology of cockroaches. We also learn the more serious life on a combat-ready ship with the recounting of the August 4, 1964 "Tonkin Gulf Incident," and subsequent rendering of events that lead to the U.S. Congress resolution authorizing armed forces to be committed to what was to become the Vietnam War.

In early chapters, McDaniel relates that as a childhood and lifelong condition, he lacks depth perception, essentially seeing space and distance as flat. One can read into this that his persistence to overcome this limitation developed in him a defining need for grounding in physical evidence, what one can see. The disability became an advantage, evident in the training he endured to be a successful marathon competitor, and more so as a scientist whose one goal is truthfulness and accurate recording. He writes in the Prelude: "Articulating on paper what I thought to be true clarified inconsistencies and forced verification." This helps explain his concluding sentence, "the value and importance of a worldview grounded in verifiable evidence and reality-based agreement," the central theme of *"Beauty Won Me Over: A Scientist's Life."*

McDaniel's work is not explained solely by devotion to the discipline of science. There is a moral compass here that comes from his early life, people who became mentors, and his own internal gyroscope. And there is his recall of lines from the movie, *How the West Was Won,* a lyrical mnemonic on how to endure life challenges: "*Crack the whip, draw some blood, make the leaders take the mud. Grab the wheel, heave it round, soon we'll be on hard ground.*"

A key contribution of the biography is the record of a scientist in the making, the plodding work of research, trial and error, failure, plodding on, the importance of guidance from teachers and colleagues, what Jacob Bronowski described as the "Habit of Truth," the rigorous testing and replication protocols by which scientists create a body of shared knowledge. McDaniel describes his career-long path to making what becomes his most important contribution to plant biology. He writes of the moment when it all came together. The discipline of research gained singular clarity in the image and the awe it evokes of the *Cecropia* silk moth. If truth-seeking curiosity is the discipline, beauty is its inspiration.

Neither persistence and moral compass, nor accomplishment in scientific endeavour, fully describe McDaniel's life. There are teaching and leadership accomplishments described with modesty, but all the same, remarkable. He describes contributions to interactive learning at Rensselaer Polytechnic Institute, a trend to replace "lecture dominated teaching" with self-directed learning, encouraging students to ask and answer their own questions. The trend spread across universities and schools in the 1990s with the advent of affordable computers and web resources, including videos that made science and humanities accessible worldwide. It took extraordinary teachers with the capacity to engage beyond one's own discipline to lead such innovations. The author was one of them.

I leave for the reader to discovery many other examples: from science newsletters, conferences, teaching at universities with distinct approaches to student honour codes, creation of an "energy positive" home, vegetable gardens and tall grass prairie, organising the building of an affordable house, facilitating Oberlin's first Climate Action Plan in 2012 to stewarding several million dollars from the sale of Renewable Energy Certificates with a focus on implementing weatherisation and energy efficiency for those houses in most need.

Mary McDaniel's contributions to teaching are embodied in an extraordinary account: In 1970, pregnant with their first child, Mary successfully fought against a ruling of Middletown CT school board that she be fired after the fifth month of pregnancy from her kindergarten teaching position that helped break the barrier of prejudice against women teaching during pregnancy. She taught until the birth and four days later brought her son—now a scientist!—to her classroom. The case was cited by U.S. Supreme Court when establishing rights of women to teach while pregnant.

Beauty Won Me Over: A Scientist's Life is the latest of Carl McDaniel's books, each one an important contribution to "a worldview grounded in verifiable evidence and reality-based agreement," that is, lessons that help prepare for the future. Readers will find their own favourite passages that express how we can live within the truth of ecological principles. I found mine towards the end of the discussion of science, biology and evolution in Chapter 6.

These processes and the principles that undergird them are fundamental to humanity's survival because they enable an understanding of what we are, where we came from, and our relationship with the rest of life. Humans are just one species evolved over a vast period of time and among millions that make up the ever-changing fabric of biodiversity on which humanity and all forms of life utterly depend.

Don Watson, Fellow American Institute of Architects, Certified Interpretive Planner, and Principal of EarthRise design

Prelude

The second of three children, I was born to a middle-class, Caucasian family in the United States as the dark shadow of World War II fell across our nation and world. Now, when I look back and into the future, it is clear that many in my generation—certainly not all—have lived during the sweet spot of the 20th century, if not in all humanity's existence. We were blessed with myriad opportunities that few others have ever had.

My eyes were severely crossed at birth. Other people saw the whites of my eyes while I saw darkness. Remarkably, I did not live a sightless life. Mid-20th-century eye surgery enabled me to see at just nine months old.

I attended public schools and participated in sports—baseball, basketball, cross country, football, swimming, and track—while also enjoying canoeing, fishing, ice skating, tennis, and wandering through patches of yet-to-be-developed land near our home in northern New Jersey. Although surgery as an infant enabled me to see, I could not effectively fuse images. Thus, my lack of depth perception led to poor performance in certain sports. Running became my sport. Unlike most other amateur athletics, distance running required no particular talent, just mental toughness and a capacity to push limits.

My eventual success as a runner, and running itself, were important to my maturation. The 'race' became a metaphor for existence. Running taught me that the essence of living life well, and to its fullest, is giving body and soul to all one does.

As far back as I can remember, I was serious and committed to whatever I was doing—perhaps to a fault. Prior to college, I had had no meaningful acquaintance with academic scientists, yet this is what I became, and my worldview is grounded in evidence-based reality. Why am I the person I am? What were the forces of nature and nurture that shaped me?

Over the years, the discipline of writing on scientific and other topics enabled me to realise that many, if not most, of the initial ideas and conclusions in my head were off the mark. Articulating in words on paper what I thought to be true clarified inconsistencies and forced verification.

What follows is the story of my life. I write to understand why and how I became a scientist. For the reader, I write to provide an appreciation for what ordinary scientists do as well as to convey the value and importance of a worldview grounded in verifiable-evidence and reality-based agreement.

Chapter 1
Is There a Scientist in Me?

Dale Bardo knew he wanted to be a physician when he entered Oberlin College. I had no idea what I wanted my future to be. During my senior year of high school, two friends and I would meet Sunday afternoons to consider the writings of Plato, Aristotle, and John Locke. Despite 'the blind leading the blind' nature of these get-togethers, they inspired me to consider a philosophy major in college. During my first semester at Oberlin, I took introduction to philosophy, which turned out to be concerned with the meaning of words, logic, and esoteric concepts. No longer inspired, I concluded that philosophy was not for me. I needed something more concrete, more grounded in what I could see and touch.

I had fond memories of my high school chemistry teacher, Mr Voltz. He walked with a limp due to a metal plate in one leg—the result of a war injury, I presume—and he had a no-nonsense approach to students. "I am not here to tie your shoes or wipe your snotty noses," he announced on the first day of class. "I expect you to be in your seats when the bell rings, ready to learn." With little more introduction, he started talking chemistry. When the period ended, he warned, "We will have unannounced quizzes."

At the start of our second class, Voltz stated, "Take out a piece of paper," followed immediately by, "First question: What is the chemical formula for water?" After several seconds, "Second question…" None of us had yet to take out our notebooks, much less sheets of paper for quiz answers. Thereafter, I would rush to my seat before each class and have prepared an answer sheet numbered one through ten. I liked chemistry and earned As. I also earned As in biology and physics, but Voltz had left a particular impression. I took chemistry during my freshman year at Oberlin and did well, so I formally declared it my major.

*****

I began thinking about college during Christmas break of my senior year in high school. We lived in Wauwatosa, Wisconsin, and my brother, Hugh, was attending Texas A&M, my dad's alma mater. But Texas A&M did not fit my three criteria for college: coed, no fraternities, and a predominantly academic atmosphere. Many of my brother's colleagues were home for Christmas, as were older siblings of friends. I asked for suggestions and settled on Beloit College, Colorado College, Oberlin College, and Stanford University. My dad favoured

Rice University, perhaps because it charged no tuition, so I applied there, too. I had never heard of Oberlin, but the older sister of one of my philosophy buddies had flatly stated, "Oberlin is your school." I was accepted at Beloit, Colorado, and Oberlin and was waitlisted at Stanford. I chose Oberlin because it was the school that met best my three criteria.

I handled the three semesters of calculus required by the chemistry major—the first two at Oberlin and the third in summer school at University of Wisconsin-Milwaukee. I pushed myself hard and liked calculus. My grades, C+, B, and A, were instructive on the academic rigor at these two schools. At Oberlin I ranked in the middle of the class while at the University of Wisconsin-Milwaukee, I ranked way above the median. Oberlin had both higher standards and more academically talented students: on May 26, 1957, the *Chicago Tribune* rated it as the best liberal arts college in the country in a ranking of the country's higher educational institutions. A high school friend had warned me about academics at Oberlin; she would never go to Oberlin because everybody gets some Cs. She was very bright and knew her academic interest—writing. All I knew was that I was expected to go to college.

I earned one of my few B^+s in introductory chemistry and did fine in organic and quantitative chemistry—all Bs. By the second semester of my sophomore year, I had come to grips with what junior year meant for Oberlin chemistry majors: physical chemistry. My experience with calculus at Oberlin gave me pause. Physical chemistry was taught by Norm Craig and Richard Schoonmaker, both newly minted PhDs who were very demanding. Success required a brain capable of simultaneously managing calculus, physics, and chemistry all at advanced levels. It was likely the most challenging course on campus. Students spent tens of hours—as much as 50—a week in class, lab, and studying. Although I had enjoyed and earned As in high school physics, my gut told me that taking college physics and physical chemistry at the same time spelled disaster. I needed to change my major, again.

In truth, the decision to become a chemistry-biology major was easy for two reasons. First, while majoring in chemistry, I had taken many of the courses that were required for the chemistry-biology major. Second, I liked plants and being outdoors. Growing up in Demarest, New Jersey, my best friend in early grade school had been the girl next door, Nelly Farnum. Our yards shared a boundary. She was athletic and adventuresome. She could catch barehanded the fastest hardball I could throw. We wandered the wooded areas near our homes just being curious, but also in search of wild asparagus, ferns, violets, and other plants to move to our yards. Meeting my dad as he walked home from the train on which he commuted to New York City, we would tell him of our discoveries and ask for his help transplanting some of them. I had my own garden in a vacant lot adjoining our property and loved growing things: gourds, okra, cotton, and, of course, asparagus. One day we found a handsome 4-foot-tall apple tree that was about to be sacrificed for a new house site. The next Saturday my dad helped us move it to the hedgerow next to my garden. When we moved to Wauwatosa a half dozen years later, the tree was 20 feet tall and beautiful.

Nelly's family moved to a neighbouring town when she was in fourth grade, and our lives went their separate ways. About eight years ago I received an email from the address farnum@cornell.edu asking me if I had lived in Demarest. I responded, "Are you the girl who could catch a hardball barehanded?" She was!

After 60 years, we met at her home near Ithaca, New York. She had just retired from a professorship at Cornell University after a career studying cellular activity associated with long bone growth. Her beautiful gardens were filled with diverse plants that echoed our joint adventures in the woods. During our visit I learned that, after graduating from Stanford, Nelly had been a Peace Corps volunteer in the Kingdom of Tonga where she taught English and math in middle school for two years and then spent a third year as Rodent Control Officer—rodents damage coconuts. This and her avid engagement against hydrological fracking for the recovery of oil and natural gas affirmed her lifetime commitment to fostering a sane society. Our early shared interests in life and the outdoors had led us to similar accomplishments and concerns for a better world.

My decision to change majors at Oberlin to chemistry-biology soon proved wise. First, physics itself was difficult enough. A friend from freshman year and I spent hours studying physics together. He quickly grasped the concepts and how to solve problems and then learned even more by guiding me through the material when I got stumped. Our physics professor soon recognised that the subject did not come easily to me. When my friend's tutoring was insufficient, the professor did his best to help me understand and solve homework problems. I passed two semesters of physics with a C+ and a B-. Second, I believed then, and still do, that Oberlin's physical-chemistry curriculum was too steep a hill for me to climb. I did take a biologically oriented physical chemistry course in graduate school, but the expectations and focus of the course were different from Oberlin's, and I had matured intellectually.

By the time I became a chemistry-biology major, Dale Bardo and I had become good friends. Dale's family lived in Lusk, Wyoming, just west of the Nebraska border and 150 miles north of Cheyenne. I was romantic about the American West then. In the late 1950s, *How the West Was Won* was my favourite movie—it captured my youthful imagination: wide-open spaces, buffalo herds filling the ground to the horizon, wagon trains, transcontinental railways, physical hardships, promises of unbounded adventure, the magic of new places, and taming the wilderness in service to a better life. I still remember a few lines from the soundtrack: "… crack the whip, draw some blood, make the leaders take the mud. Grab the wheel, heave it round, soon we'll be on hard ground." The country's western expansion was underway. I had visited Texas when I was 10 to meet my parents' relatives—my dad grew up in Hillsboro and my mom in San Angelo. Each had moved to New York City in the 1930s, where they met. Perhaps that Texas experience set the stage for my yearnings to go west that would soon be realised.

In 1962 I took calculus and French in summer school and worked a summer job at the Wauwatosa public library. Dale was working for his dad at the *Lusk Herald*, the town's newspaper that his family owned and managed. Dale arranged for me to spend the month of August working as a construction labourer for Frontier Lumber Company. I took the train from Chicago, riding in the dome car and enjoying vistas that changed from forest and rolling fields to mile-high short-grass prairie. Dale picked me up that afternoon in a friend's red and white convertible. Talk about an adrenalin high! Dale is typically soft-spoken and conservative with words, but with the top down, he and I talked enthusiastically, and 150 miles flew by in well under two hours.

Wauwatosa, Wisconsin, is flat compared to Demarest, New Jersey. Big Sky Country, Wyoming, however, is in a totally different class of flatness—I had yet to ride the train across Australia's Nullabor Plain that extends hundreds of miles with less than a foot difference in elevation—the ultimate in flatness. A 90-year-old Lusk native who lived in the Bardos' backroom told stories, one of which I remember quite well. Long before paved roads came to town and when horse travel was common, he said that he dropped his wallet on the main street one evening. Realising the loss, he retraced his steps the next morning, finding his wallet in the street where it had dropped. I loved that character trait of western people and was seduced by mile-high, Big Sky Country.

I had never done construction, but did physical work—gardening, cutting firewood, shovelling snow, and participating in all the standard sports. I was in excellent physical condition from having run the mile and two-mile races on Oberlin's track team. On my first day at Frontier Lumber Company, I had but one objective: survival. The men I would be joining were labourers and craftsmen; my job for one month was their life's work.

The foreman, stocky and just under six feet tall with blond hair, went by the name Whitie. He sent me with a crew to pour a cement floor in a creamery basement. I was to shovel the concrete as it was poured into areas of the basement not reached by the cement truck shoot. Shovelling concrete is exhausting work. The morning break did not come too soon. By lunchtime, as I ate my sandwich and fruit, I seriously doubted that I could continue at the same pace for five more hours. Fortunately, Whitie came by as we finished lunch and moved me to another job: carrying cinder blocks into a newly excavated basement. Perhaps I had passed my newcomer trial. In any case, I did this and all jobs without complaint. I slept well, and for all four weeks my hands were so stiff each morning that it took half an hour before I could tie my bootlaces.

Dale and I returned to Oberlin, Dale for his senior year and medical school applications, I for my junior year. Almost everybody at Oberlin—students, staff, and faculty—had a bicycle back then, as students were prohibited from bringing cars to campus. We rode bikes everywhere, regardless of weather. I enjoyed bicycling in the open country that was less than a mile from campus and even to Lake Erie, just a 40-minute ride north. Over the fall term, an idea emerged in Dale's mind and mine. Next summer…we would ride bicycles from coast to coast! We wrote to Schwinn and Huffy for sponsorship. A polite no came from Schwinn. Huffy offered two balloon-tire bikes that were not any better than the bikes we had. In the early 1960s, few people were riding hundreds of miles, almost none thousands. I am sure touring bikes must have been available, but we found none. Lacking a better choice, we purchased aluminium Italian racing bikes and began training in the spring. We figured that we would take temporary jobs along the way to cover costs. We had limited financial resources, and helping with Oberlin's tuition and room already stretched our parents' resources—we both had dining-hall jobs that covered board.

To our Frejus bicycles we added back wheel racks with side packs that held 30 pounds of clothes and gear that included a tent, sleeping bags, simple tools, and spare tires. We sewed day-glow patches onto our ball caps and side packs. We had Oberlin friends along our route with whom we could stay.

Wheeling out of Oberlin at 6:30 a.m. on June 18, 1963, we left the van der Pyle Hilton, home of Mr and Mrs van der Pyle—Van and Betty, where Dale and I had enjoyed many fun Oberlin times. We headed east to Bill Saint's home in eastern Pennsylvania, 160 miles away and a two-day ride.

1.1: Dale Bardo and author with bikes.

We sought work on both days, once to help with haying and again at a produce farm, with no luck. Bill Saint's family ran a nursery. They gave us two days' work that netted $4.77 plus food. Our trip-log entry the day we left the Saint's home stated, "After breakfast we decided, because of the time involved and the small amount gained, to eliminate for the most part, the working-as-we-go aspect of the trip." We had our first reality check.

Before passing through the Adirondack Mountains, we had had three blowouts and had run out of spare tires. Our second reality check: a thin racing tire would not support a rider and 30 pounds of gear. We laid over for two days at Moffitt Beach State Park while waiting for more tires to reach the post office in Speculator, New York. We rested and had time to reflect on the trip and our relationship. Both Dale and I were bothered. Something was missing. For well over a year at Oberlin and in Lusk we had done everything together, not as separate individuals. But now we were not clicking as we had expected. We were not sure what we wanted from this bicycling adventure, nor its value. However, we were serious about life and were committed to the decisions we had made. Well-rested with spare tires packed, we headed for Camp Norway in Ely, Vermont, where I had been a camper and counsellor for three summers in high school. From there we went to camp Pemigewasset in New Hampshire's White Mountains, where Bill Saint and Johnny van der Pyle, son of our Oberlin hosts Betty and Van, were employed. Then, we peddled on to Boston.

Dale and I were in fantastic condition and felt more comfortable riding than walking. Nevertheless, we had a few hellish, 100-degree days and several exhausting mountain ascents. The ride to Boston was a hard one: 130 miles, 12 hours, and 100 degrees. We had covered 924 miles since leaving Oberlin.

The next morning, after staying with a fellow Obie, we toured Boston and spent half an hour walking around Harvard Square. By 1 p.m., we were on the road. Again, it was very hot, and our spirits flagged. We considered taking a train back to Ohio, but after an hour's talk and rest, we decided to keep peddling. We ate supper overlooking the Mayflower II. That evening we had a hard time finding a place to bed down. Eventually a family campground owner, after originally denying us a site, called us back.

The hard ride into Boston and another 100-degree day caused the concerns we had had at Moffitt Beach four days earlier to re-emerge. A change was needed. We decided to skip Cape Cod and headed directly to Providence, Rhode Island. We stayed with Mildred Mosier, a retired schoolteacher and Baptist missionary I knew. She fed us dinner and gave us a place to sleep. Dale rose at 1:40 a.m. to catch a bus to Pennington, New Jersey, to visit his steady girlfriend, Mickey Himes. I left four hours later with plans to meet Dale at Mickey's house in five days. I headed for West Haven, Connecticut, to visit Bill Frazier. He was a year behind me at Oberlin and worked for me in the dining halls, where I was headwaiter.

It was July 4, 1963 and I wanted to get to West Haven before traffic got heavy. Not thinking about topography, I headed west on U.S. Route 6, which runs across the middle of the state. Thousands of years earlier, glaciers had moved south, gouging out the land, and I was soon climbing steep hill after steep hill. When heading down, I freewheeled, reaching speeds of 40 to 50 mph. The two-lane road had no shoulder, but in many places had cables on posts to prevent cars from going off the sides. About midmorning, going down a long hill, a line of four or five cars was coming up ahead of me. Suddenly, a sports car at the rear of the line pulled out, passing the other cars at perhaps 60 mph. We were closing at about 100 mph. With over 1,000 miles under my belt, I was a skilled biker. I eased over to the edge of the road. The sports car flashed by me, its side mirror almost touching the mirror on my handlebars. At that instant, the driver saw me, and I him; horror was in his eyes and on his face. My heart rate raced with a huge surge of adrenalin. It took a long time for me to calm down; I had come as close to death then as I ever have.

I reached Middletown late that morning and called Bill. His dad answered and gave me directions for what should have been a 40-minute ride. I took Route

15 as Mr Frazier had suggested—a four-lane parkway with gently graded slopes that allowed me to set a steady pace and take my own lane, leaving the other lane for cars. No sooner had I begun to enjoy the ride, than a police car pulled me over. The officer told me that bikes were not permitted on the parkway and took me to the next exit. My map revealed no easy route into New Haven. Three hours later, after being misdirected three or four times, I reached Bill's driveway. I had had little to drink and nothing to eat since leaving Providence. As I stepped off my bike, Bill's sister, Mary, burst out of the back door saying, "We have been waiting for you! Where have you been?"

I replied, "Trying to find your house. Nobody gives good directions."

She responded, "Are you hungry?"

"Yes!" After eating a whole apple pie and half a gallon of ice cream, I felt much better. Little did I know where this would lead.

I spent the evening of July 4 with Bill, his girlfriend Mary Thurston, his sister Mary, and several of her friends on their front lawn, listening to a band concert in the town green across the street. Bill's dad, Reverend Donald Frazier, had warned us not to mess with the town toughs who certainly would be out on the green. As it turned out, the 'toughs' were his daughter Mary and her three girlfriends. After they forcefully took my shoes, the game became tackling and trying to hold me down. My high school days were not like that, but I had no complaints. When the concert was over, these high-spirited girls took me swimming in Long Island Sound, where I nearly froze. I slept in the next morning and spent the rest of the day leisurely, swimming with Mary and touring Yale with her parents. I felt very much at home with Bill's family.

After breakfast, Bill drove me 35 miles to Danbury. From there I peddled to Camp Sharparoon, where my sister was a summer counsellor. That evening at sunset I crossed the Hudson River at Bear Mountain and slept in the bushes off the side of the road. In the morning, I biked to Demarest (where I had lived as a child), ate breakfast with family friends, went to church, and continued visiting. That evening, I visited with a girl I knew who was then a sophomore at Wellesley College, talking mostly about her impressions of freshman year. Wellesley women's social lives and relationships dominated the conversation—they were very different from women at Oberlin, who were not only engaged with their lives, but also with studies and societal issues of the day.

Monday was rainy as I headed for Pennington, where broken glass on the road ruined my back tire. My notes on the day included, "The traffic and roads were *hell*, and guess I was lucky to make it." I reunited with Dale late that afternoon as he was finishing up the day's painting on the Himes's house. Two days later, we saddled up and hit the road. Four miles out, boom! We hitched back to Princeton, where we had changed to clincher rim tires and tubes, to get the faulty tube replaced. We rolled another four miles, and the same tire blew.

We put a spare tube in the new tire and had a long talk. We were mentally exhausted, to the point that we called Greyhound in Philadelphia to see when the next bus left for Cleveland. But on reflection, we knew quitting was not an option; we had to ride on. After five days, we were 130 miles from Oberlin. After sleeping on the gravel parking lot of the restaurant where we had had dinner—we could have slept on anything—we mounted up. It was 6:20 a.m. Arriving in Oberlin at 7:15 p.m. we checked out the town: Tappan Square, Hall Auditorium, Finney Chapel, Carnegie Library, and Peters, Talcott, South, and Johnson House. Our tour ended at 371 West College Street, the van der Pyle Hilton, where we had started out 29 days and 1,861 miles earlier. Although our goal of bicycling to Boston and then across the country to California was not realised, Dale and I gained something more: a deeper understanding of life, people, each other, and ourselves. And we experienced the north eastern part of the country and some of its people close-up.

As senior year progressed, I thought more and more about what to do after graduation. I was clearly interested in biology and had initiated a senior research project in plant physiology with Tom Scott, a newly appointed assistant professor. Elizabeth Gilbert, also an assistant professor in biology, was my major academic advisor. Naturally, it was from these two faculty members that I sought advice. Because almost all Oberlin graduates went on to advanced degrees, I, too, was contemplating graduate studies. Many in the sciences at Oberlin, like Dale, went to medical school. I had no interest in becoming a physician; besides, my grades were insufficient to get medical school interviews. Beyond that, I was naïve about options.

Gilbert's assessment of my potential for graduate studies in biology was minimal; however, she believed I would do well teaching high school science. Scott concurred. I did have a general interest in teaching—I had taught in Boy Scouts and at summer camps. I also contemplated institutional food service. I had enjoyed working for more than three years for Saga, the company contracted to provide Oberlin College's food service, as the headwaiter in four dining halls. The top Saga person at Oberlin offered to sponsor me for a manager position.

Dale was now studying at Case Western Reserve medical school in Cleveland. On his rare visits to Oberlin we discussed my quandary. We ran through the possibilities. He adamantly stated, "Carl, you would be bored stiff after a few years of high school teaching. You need bigger challenges." Dale knew me better than I did myself. Perhaps he saw the scientist I had yet to discover.

I began my graduate school applications with a winter application to the department of biology at the University of California, Berkeley, but my grades did not meet the school's minimum requirement. I then applied to a master's program in botany at the University of Michigan—not appreciating that those major research universities wanted PhD students—and to the biology department at Wesleyan University in Middletown, Connecticut, which at the time offered only master's degrees. Michigan accepted me with no financial assistance while Wesleyan offered a research fellowship that provided $1,800 annually in addition to tuition and fees. I took it.

After graduating from Oberlin, I headed back to Wyoming for summer work at Frontier Lumber, this time as mud boy for Alex Miller, a mason and a German immigrant. He was my height and wiry, a no-nonsense craftsman who mostly kept his own counsel. I drove a pickup truck hauling tools, bags of cement, sand, and whatever else Miller needed. He had high standards and expected me to learn to anticipate his needs. If he had to call me for mud (mortar), brick, block or stone, I felt that I was falling short; however, Miller never said so. I came to understand that he disliked being idle and did not wish to waste time.

1.2: Author next to the Frontier Lumber Company truck he drove.

Sometimes we had jobs in neighbouring towns that meant traveling 50 or more miles before a 10-hour workday. I would meet Miller at the lumber yard, and he would drive. Wyoming had no speed limit then, so 100 mph was his standard. On the job, Miller did not say much, but during these drives he spoke about the prejudice he and his family had experienced during and after World War II as a German in the U.S. I sensed it was still there for him, even two decades later. The reality of prejudice was not new to me. Oberlin colleagues went south to register voters, and my father, like many southerners then, considered Negroes inferior to whites. However, this was the first time that a person I knew and respected had experienced this type of deep cultural isolation.

I again lived with Dale's family that summer; this time he too worked for Frontier Lumber. After work, he and I painted the Bardo house. It made for a long day—up at 6, to bed after 10—but we had a grand time walking to and from work and then talking with his parents over dinner and with each other while painting.

I did not appreciate the volume of a railroad boxcar until Dale and I unloaded one for Frontier Lumber over a weekend. We hauled the lumber on a good-size flatbed truck from the boxcar to the lumberyard. Not fully grasping the practical implications of inertia, I took a turn too fast. The huge load of lumber shifted, almost tipping the truck over, with Dale riding in the back with the load. Knowledge transfer to new situations demands vigilance. This incident seared into my memory and often jumps out when I think about Dale. Good fortune saved me from terrible consequences—then and several times since when felling trees with a chainsaw and cutting wood with an axe.

At the end of the summer Dale, his sister, and I took a week to fly-fish in the Wind River Mountains. We hiked several miles up the trail and forded the river to set up our base camp on the southern side. Using branches, strips of bark, and twine we made a table on which to prepare our meals and created a stone fire circle for cooking. Each day we hiked up the river to where the depth was not much above our knees and the water sufficiently placid to fly-fish. Dale knew how to fly-fish, but this was new to me. The first day we caught nothing. Then our luck changed—perhaps our skills had improved? We ate trout for dinner and had more for breakfast. We put a line tied to a bush through the gills and placed the fish in a pool of a small stream near our site to hold them. In the morning, only the string remained; a hungry animal had enjoyed an easy catch.

This site was part of the Wind River Indian Reservation, which in the 1880s had been set aside for Eastern Shoshone and Northern Arapaho tribes. Local residents vacationed here, and the National Outdoor Leadership School (NOLS) used the area to teach outdoor living skills and consensus leadership by doing. I was so taken by the place that years later, during my family's first trip west in 1978, I took my wife Mary Frazier and our two children, Stuart and Virginia, to this place, where all cars still displayed Wyoming license plates. When our daughter was in high school, she returned to take a month long NOLS course there. She called us at the end to affirm her experience by first declaring, "I can do anything!" NOLS did well by her. She is perfectly comfortable in the woods and open spaces, which serves her well as a field botanist and fire ecologist for the U.S. Forest Service where she is expected to volunteer to fight fires or help with other major disasters.

I used my entire summer savings of $800 from working for Frontier Lumber to purchase my first car, a used 1960 blue Volkswagen bug. My dad, always an advocate of safety, looked at the tires and purchased me new ones. I drove to Wesleyan University to begin my graduate studies in biology. Dale's telling me that I needed more demanding challenges than teaching high school was the near-term reason for my being at Wesleyan. That withstanding, I came to realise deeper, more elemental reasons.

Chapter 2
Becoming a Runner

My birth in Englewood, New Jersey, on June 9, 1942, was not smooth. My umbilical cord was wrapped around my neck, preventing my mother from pushing me out. The family doctor, recognising the problem, reached in and freed me. Out I came! My mother wrote, "You had such a mop of hair that stood right up—looking as if you had a crew haircut—instantly the nurses and doctors called you Admiral." The reference was to Admiral Chester Nimitz, whose leadership as commander-in-chief of the Pacific Fleet had, four days earlier at the battle of Midway Island, prevented the Japanese from occupying Midway. The Japanese lost four heavy aircraft carriers, 332 aircraft, and hundreds of the best naval aviators in the world. Although no one realised it at the time, this was the turning point of the naval war in the Pacific.

My middle initial, N, stands for Nimitz, my mother's maiden name. She was a second cousin to the admiral, and my parents knew him personally. A letter I saved from my dad stated that Nimitz was planning to visit soon after the war, but I have no memory of it. Mary Frazier, who had become my wife, and I, however, were delighted to spend several hours with his widow, Catherine, in 1967 when the destroyer on which I served during the Vietnam War was in San Francisco.

I spent my formative childhood years in Demarest, New Jersey. Much of the town was undeveloped and provided innumerable opportunities outdoors. Many families, including mine, did not have an automobile, so we walked and rode bikes everywhere, taking the bus for more distant destinations. My dad rode the train to New York City where he worked as a sales engineer for Allis Chalmers. We lived in an established neighbourhood with vacant lots. Grade school basketball courts and baseball fields were a short, two-block walk. Natural areas of fields, woods, streams and ponds, and an abandoned sand pit were easily reached on foot or by bike.

Parenting in the 1940s and 1950s was far less intrusive than today. Yes, my parents cared deeply about us and our safety, supported and volunteered in Boy and Girl Scouts, went to parent-teacher conferences, served on the school board, and expected us to stay out of trouble. That said, their oversight of my friends and me was minimal. On weekends, holidays, and summer breaks, I had tremendous freedom. Typically, after breakfast, I would say, "Mom, I am going to Davies Pond (or some other place)." She would reply, "Be home by dinner." The age of cell phones and helicopter parenting was two generations thence. I enjoyed many hours of unsupervised time in the natural world observing, thinking, or playing alone or with other unsupervised youth.

My mother told me about a time when my brother, Hugh, at about age 5, announced that he no longer wanted to live with the family and was running away. She fixed him a sandwich, helped him pack a few things, and wished him well as she walked him to the door. That afternoon he walked back into the house and asked when dinner would be ready. I was awed by her confidence in Hugh not only to make the right decision, but also to come home safely. He never ran away again. Our daughter, Virginia, called me when her daughter Ruth was about three to ask, "What did you and Mom do to make me so self-confident and independent?" I responded, "Honestly, I do not know. Perhaps we just kept a loose safety rope on you until it was no longer needed, and you shook it off."

I do not remember having a dog when a child, but for a decade we had Freckles, a cat that delivered a litter of kittens twice a year for a total of 96 by my mother's count. We named them all—Tornado, Mississippi, Boots, Professor, Lightning, and so on. On our screened-in porch, Hugh and I kept all manner of creatures we had caught and held in jars, pails, cans, and tubs—various stages, types, and sizes of fish, tadpoles, frogs, toads, turtles, snakes, and, of course, fireflies that we brought to our bedrooms to light up the night. More than once, a praying mantis egg case kept inside for winter released many dozens of tiny mantises that we frantically collected to release outside. Apparently, a $50 fine could be imposed on a person for killing one because they were voracious insect eaters.

My world overflowed with 'wildlife'. But little did I realise then just how impoverished the biological world I had explored had become compared to what it once had been. I did not know the train that Dad rode to work or the turbines, generators, and other industrial equipment that he sold or the roads I walked to reach woods and streams were changing the fabric of life forever. Could it be otherwise? We all assess a situation in terms of our lived experiences, or at least until innocence is lost.

*****

As I noted in the prelude, running had provided the metaphor by which I understood life. Like most grade schoolboys and some girls, I played football, basketball, and baseball, and I enjoyed swimming. Baseball was my best sport, where I was better as pitcher than other positions. I misjudged fly balls and was poor at catching a football and shooting baskets. I discovered why years later in driver's education class that I had no depth perception.

Much later in graduate school I learned that the brain acquires the capacity to fuse images in the first year, thereby enabling depth perception. Development of an organism is a particular sequence of myriad events leading from egg to adult. If any event fails to occur in the proper sequence—in this case programming the brain to fuse images from two eyes to enable depth perception—it does not happen. Unlike cleft palate and other visible morphological anomalies, my lack of depth perception was not obvious to others or me. I did not know that I lacked normal depth perception. Apparently, because I did not know that I had a handicap, I persisted trying to catch fly balls and shoot baskets, believing all I needed to do was practice more and try harder. I do not remember asking my parents or anybody about my lack of skill at catching balls. Once I understood my handicap, I came to believe that my parents were not only correct to have had my eyes operated on as soon as possible, but also wise for not giving me an excuse to slack off.

My dad had learned to swim when his older brothers grabbed him by the arms and legs and tossed him into their swimming hole, saying, "Hugh, time to learn to swim." Though he never liked swimming, my siblings and I were taught to swim at the YMCA. I had lessons on Saturdays. Sometimes the Y held an informal Saturday swim meet. I remember clearly one of those meets, maybe because I was told about it afterward. I must have been about 10. Several of the bigger boys were captains and picked swimmers to be on their teams. Small, skinny, and shivering, I was picked last. I was fourth to swim in a 200-yard relay in which each boy swam 50 yards. My team was in last place when I dove into the water. Soon everybody was yelling and cheering, because I was passing other swimmers and gaining on the leader. Inspired by the encouragement, I swam faster. We won! I was picked out of the water and hoisted up onto somebody's shoulders. This victory is memorable because it was the exceptional outcome for me as a youngster in sports.

2.1: Author standing in lake holding turtle at age 14.

Demarest, like other small towns, had a Memorial Day parade that included Boy Scouts, Girl Scouts, the American Legion, fire trucks, and other groups. It was followed by a remembrance ceremony for those who had died in past wars, which concluded with a 21-gun salute and the playing of taps. Afterward came food, rides on fire trucks, games, and foot races. My siblings and I took part, as did our friends and others. Of course, I never did win although I wanted to win. In our neighbourhood, we routinely had pick-up races around the block, even imposing silly conditions like not breathing through our mouths. Lacking depth perception was not a handicap. Sometimes I would win or run the fastest time.

My mom was down to earth and a hard worker. Having taken care of her mother as she died and living through the great depression, she knew how to make ends meet with modest resources. She was a bargain hunter—furnishing the house with junkshop and Goodwill furniture and household items. In addition, she glued, upholstered, and refinished the furniture as well as painted and wallpapered the rooms, some more than once. I still have the pedal-operated sewing machine she used to make and repair our clothes, including numerous Halloween costumes. Mom created basic but tasty meals from what her budget allowed, along with cookies, pies, cakes, and many flavours of fudge. We never went hungry, nor did we go out to eat until I was in high school, and then infrequently and not fancy. Like most of our neighbours and the families of my friends, we were middle class with sufficient funds for food and clothes, eventually a house, and then a car. The house and car came because my dad was hard working and fiscally conservative. Like my mom, he weathered the depression. After graduating from Texas A&M, he worked for Allis Chalmers in Milwaukee. In 1931 he was let go. Rather than stand in a bread line, he went south to Gary, Indiana, where he got a job reading oil tank levels: 12 hours a day, six days a week. The pay was food, a place to sleep, and 50 cents a day.

2.2: Author's parents.

More than a decade after working in Gary, Indiana, my dad's character traits impressed one of his more senior male co-workers after returning to Allis Chalmers. Not long after my sister was born, he said to my dad: "Hugh, you have a family now. You are renting a good house in a fine community. See if you can buy the house. I will provide the down payment. I know you are good for the loan. Just pay me back as you can." Half a dozen years later the man made the same arrangement for a used car. In 1952, the car enabled our first family vacation, a trip to Texas to visit relatives.

My parents' frugal approach to living, along with their responsible engagement with community and sincere ethical commitment regarding honesty and the repayment of debt, instructed me well to be an honest, community-oriented, fiscal conservative. Nevertheless, over the past few decades I have come to believe that Mildred Mosier, a teacher and then a Baptist missionary who grew up in Demarest, also honed my values and approach to living. After serving 20 years in Burma (now Myanmar), she retired to live in her family's farmhouse near our home.

All the children in our neighbourhood in the late 1940s and 1950s knew Mosier. Her front yard was graced with large maple trees that dropped tens of thousands of seeds, many of which germinated into seedlings. Preteen children were invited to pull maple seedlings from her yard, earning five cents per hundred. Mosier, having been an elementary school teacher, used this exercise to teach simple math. She also sought to identify a responsible boy to become her helper; he needed to be 12 and strong enough to manage the tasks she required to be done. Interested children were instructed to put piles of seedlings in neat rows of 10. I passed muster, and at 12 in 1954, became her boy helper.

I learned from Mosier how to correctly clean toilets, sinks, the kitchen, bathroom and hardwood floors, hand vacuum and rotate oriental rugs, dust woodwork and various oriental souvenirs, wash windows, put up and take down storm windows, rake and compost leaves, manage and harvest a vegetable garden, and use tools like a mattock to dig out weeds or other unwanted plants. She was organised—everything had its place—and frugal. Mosier saved string, aluminium foil, wrapping paper, and glass jars for reuse. Although she had a car, she rode her bicycle everywhere, even two or three miles to neighbouring towns. If a pound of hamburger was a few cents less in Tenafly, she would ride the three miles there. I was taught by doing and by example, as well as by the stories she would tell as we worked. Over time I developed a deep respect for her and how she lived. Even though we moved to Wisconsin in 1956, she and I stayed in touch. Dale and I stayed with her in Providence, Rhode Island, on our bike trip, and in 1967, Mary and I invited her to our wedding. Mosier represented the grandparents I never knew.

2.3: Mildred Mosier on the far right in front of the author along with Mary Frazier McDaniel and her maternal grandmother Minnie Hart Frazier on the left at author's wedding.

My mother said more than once, "Idle hands and minds create havoc in one's life." The summer of 1956, when our family moved to Wauwatosa, Wisconsin, Hugh and I went to Camp Norway on Lake Fairlee in Vermont for two months. I was a camper, Hugh a junior counsellor who earned $50 for the summer. I had mixed feelings when we boarded the train in Grand Central Station in New York bound for Vermont and my first extended period away from home. I learned to play tennis, shoot a 22-caliber rifle, canoe, water ski, and to improve my swimming and archery skills. I loved being in the open air—we slept in structures that had wooden floors, half walls, and a roof. I adjusted easily and did well, earning the best camper award in the senior division that entitled me to return the next summer for free as a counsellor in training.

Lake Fairlee, in addition to Camp Norway, had two camps for girls, and from time to time the girls would walk around the lake. We held dances with the camps, and each of us got to know a few girls. On the mornings when the girls walked, some of us got permission to fish off Norway's swimming dock, where we would meet up with a girl who we had met at a dance. One of these mornings, a guy was watching the girls so intensely that he stepped backwards into the water with tackle box and pole in his hands. We laughed so loudly that the girls all looked to see what was so funny—a most memorable event of teenage girls distracting a boy. The girls walking around the lake motivated some Norway campers to organise a run around the lake. I joined them for my first long distance run-walk, eight miles in 75 minutes, a pace just over nine minutes per mile.

At the end of camp, Hugh and I flew to our new home in Wauwatosa, our first airplane flight. Hugh enrolled as a junior at Wauwatosa Senior High School and I as a ninth grader at Longfellow Middle School. He went out for cross country, I for the swim team. I did not like indoor swimming. The chlorine hurt my eyes, and water got in my ears, causing infections. The next year, I joined my brother in high school and on the cross-country team. Later, while in driver's education class, I learned that no matter how long and hard I tried, I would never improve much at sports that relied heavily on depth perception. Running became my sport. It gave me identity and colleagues for the better part of 50 years.

*****

Cross country is scored atypically—the team with the lowest score wins. Runners are assigned points equal to their finishing places, but only points earned by a team's first five finishers are the ones that count. For example, in a dual meet with 16 runners the finishing positions could be as shown in table below.

2.4: Table of possible scoring for a cross country meet.

Team A		Team B	
Place	Points	Place	Points
1	1	3	3
2	2	4	4
6	6	5	5
9	9	7	7
13	13	8	8
14	--	10	--
15	--	11	--
16	--	12	--
Total Points	31	Total Points	27

Team points are: A team, 31; B team, 27. Thus, B won. The sixth, seventh, and eighth runners of team B are called 'also-rans' because they did not contribute to their team's score; nevertheless, they won the meet for their team by displacing the other teams' runners to higher numbered positions. Cross country is really a team sport in this mostly unappreciated way.

*****

Wauwatosa High School had rules for student athletes: demonstrate sportsmanlike conduct on and off the field, follow curfew (10:30 on school nights and 12:30 on weekends), and, of course, avoid tobacco and alcohol. I was pleased to have rules for tobacco and alcohol use. These rules provided a valid reason for not indulging thereby avoided peer pressure to do so. More relevant, I did not want to indulge—I had observed their consequences.

From my earliest memories, my dad smoked. He had had a lot of stress as a sales engineer at Allis Chalmers, where he was hired after the 1930s depression, negotiating million-dollar contracts for turbines, generators, motors, and other heavy equipment needed by industry and the military. He would cough and hack so loudly each morning that he sounded horribly ill. I decided then to never smoke. These episodes got so bad that he quit smoking for several months. Then during a day of intense negotiations, one of his cohorts who smoked looked at him, asking, "Mac, need a cigarette?" and offered one from his pack. That is all it took—stress and the need for an adrenaline surge to sharpen his brain. Soon he was back to smoking a pack or more a day. This was repeated three more times over the next 10 years until in 1955 he quit for good.

Equally seminal to my teenage life as my dad's morning hacking was an incident that occurred during a pool party at the home of an Allis Chalmers vice president. I was not yet a skilled swimmer. Someone shouted, "There is a kid at the bottom of the pool!" My dad dove in and pulled me out, sputtering water and frightened. It is possible that he had saved my life. When the party broke up, we got into the car and headed home. Dad was drunk and got lost, ending up in a railroad yard and almost crashing into a boxcar. He was disoriented, erratically crossing tracks only to come to one dead end after another. My parents were arguing loudly while my siblings and I sat in the back seat, scared to death. I resolved that night that if this were where alcohol leads, I would never drink. 'Never' lasted 15 years, until 1968 when I was a line officer on the Richard B. Anderson DD786.

Wauwatosa's cross-country coach was Don A. Gehrmann, a world-famous middle distant runner. From 1949 to 1952 he was unbeatable winning 39 consecutive mile races against the best runners in the world. When at the University of Wisconsin, his coach considered him the best runner the world had ever seen. At Madison Square Garden he was given a standing ovation when running his warm-up lap. People familiar with his running believed, as did Gehrmann himself, that he had the ability to run a sub four-minute mile before Roger Bannister did it in 1954. Gehrmann ran to win, never for time[1].

Gehrmann was in charge of the cross country and track programs for 13 years, beginning in 1954, at Wauwatosa High School, popularly known as Tosa. Having been a champion runner his entire career, his primary goals as coach were to work with all runners to help them achieve their potential and to groom championship teams.

Several weeks before school started in 1957, Gehrmann sent each prospective runner a letter that stated:

I expect that you are getting yourself in proper condition in order to represent Wauwatosa High School as potential Suburban Cross-Country Champions in 1957.

---

[1] Gary D'Amato of the Journal Sentinel, http://archive.jsonline.com/sports/etc/olympian-runner-don-gehrmann-looks-back-at-his-accomplishments-fb67d8j-163623646.html/ accessed 9 January 2018

I plan on training right with you and want you to have all the soreness and stiffness out of the way before we start working out as a team.

We will begin our training as a team on Tuesday, September 3. I shall look forward to seeing all of you at 4:00 P.M. at the Senior High locker room. (Instructions for training and other information.)

I look forward to a very successful and enjoyable season. If you want to be on a championship team, act like a champion, compete as a champion, and you will be a champion.

I had spent the summer of 1957 at Camp Norway as a Counsellor in Training (CIT). I earned recognition as the best CIT, the highest Camp Norway award, allowing me to return the next summer as a junior counsellor.

Although camp life put me in fine physical condition, I was unprepared for the running regime prescribed in Gehrmann's letter. Needless to say, I was still sore and stiff when team practice began. Over the next two months, I got myself in excellent physical condition for the first time and learned how good one feels in that state.

2.5: Author after receiving at Camp Norway best CIT award.

I was not a natural runner. Gehrmann and the challenge inspired me to overcome with effort what I lacked in talent. Sometimes I ran on the A team—once beating my brother—but mostly on the B team. In one meet, Tosa's A team finished fifth against most of the suburban teams. The B team won, with me finishing third overall and a teammate first.

In the 1957 Cross Country Award Night Program, Gehrmann wrote, "Carl's improvement throughout the season was very gratifying and even surprising to many. It goes to prove what the right mental attitude can do. Carl's achievements came through hard work and the desire to win. He has a great future."

Running had become a major part of my life. I competed on the track team, running the half-mile in the spring and cross country in the fall. Seven more boys joined the team, making it 30 strong with only one senior. Competition for the 16 slots—eight on each A and B teams—grew, and a core of excellent runners emerged. More A-team meets were won where I was mostly an 'also ran'. Gehrmann's comments for me at the next year's awards night were: "Carl is a fine leader and gentleman. We are glad Carl will return next year and look for a year's maturity to make a big difference in his relaxation while running."

The following summer I worked at Camp Indian Sands in Wisconsin as a counsellor for $150. Gehrmann ran the camp, and many of the counsellors and staff were members of Tosa's track and cross-country teams. This was my fourth summer of camp experiences, and my confidence in working with people toward common and individual goals had matured. I did not fully appreciate then the efforts Gehrmann put forth to help each of us achieve our potentials as people and athletes while grooming us to be championship cross country and track teams.

In the fall, I was named co-captain of the cross-country team and was among the first five Tosa runners in most meets—no longer an 'also ran'. Our team dominated the conference, winning all but one of our conference meets. "Water, Water, Everywhere, but City Harriers Run just the Same," was the headline of a half-page *Milwaukee Journal* photo depicting a herd of runners splashing through huge puddles of water. Unlike Gehrmann, who in the 1948 Olympics ran what he considered his worst race ever in water and mud, I found them to be the perfect conditions. This was my best cross-country performance ever—I was the third runner from Tosa and ninth overall. We won with 44 points; the second team had 60. It had taken Gehrmann four years to develop a team that won Milwaukee's Suburban Cross-Country Championship.

2.6: Wauwatosa's 1959 suburban championship cross country team. Coach Gehrmann is in back row first on the left side and author is in the second row next the last on the right side.

Gehrmann's comments for me at awards night that fall were, "A born leader with a home background of fond respect and consideration for fellow men. Carl is an example to everyone. It will be impossible to replace Carl, but his magnetic desire to do his best will linger in hearts of our next year's varsity members."

*****

In moving from New Jersey to Wisconsin, I had become a leader not only in cross country and track but also in Explorer Scouts and HI-Y, the high school social group of the YMCA. In Demarest, I had participated in a Cub Scout den and a Boy Scout troop, where I learned the principles of scouting and camping that complemented our outdoor experiences with family and friends. In Wauwatosa, at the age of 14, I joined Explorer Post 21, an emergency medical post whose members were trained in first aid and which could be mobilised in 15 minutes. Each of us had to have our pack ready with water, food, and clothing for three days along with first aid gear. The post had a mobilisation phone tree. When you received the call, you called those next on the list and waited in front of your house to be picked up.

Within two years, I had earned my silver award (highest award in Explorer Scouts) and reached star but got too busy to do more. I became post secretary, senior crew leader, and chairman of the first successful district activities. When elected chairman of the Northwest District Council, I became part of the council cabinet and assumed responsibilities as chairman of several activities, including a summer beach party and Autumn Leaves and Valentine dances.

I was also very active in HI-Y, the high school boys' program of the YMCA whose charge was "to create, maintain and extend, throughout the home, school, and community, high standards of Christian character". I served two terms as the president of the HI-Y at the local YMCA as well as being chaplain of the Wauwatosa Council of HI-Ys and TRI-HIYs (girls' group). In 1960 the Wisconsin YMCA held its biennial two-day model legislature program in the Wisconsin Assembly and Senate Chambers. I was selected to be one of the three candidates for Governor and elected Governor at the pre-legislative conference held at Lawrence College in late February 1960.

In the meantime, the 1960 Wauwatosa indoor track season was going well. We had won the University of Wisconsin-Milwaukee high school indoor relays with 38 points, an almost perfect score. In those relays we won the 12-lap medley, breaking the record by 1.4 seconds, and I ran in the 20-lap relay of four half miles, which broke the record by 3.9 seconds. Gehrmann had another championship team in the making.

I told Gehrmann that I had won the election to be Youth Governor; he had written a letter of recommendation supporting me. He told me that if I attended the model legislature, I would be ineligible for the outdoor track team that spring because I would miss the Shorewood Relays, the largest event of the season. We were defending champions with a solid chance to win again. What a conundrum! I spent hours talking with my parents, fellow runners, YMCA people, and Gehrmann, who in no small way had brought out the runner in me.

Gehrmann stated his position clearly in a letter.

I have requested of Carl that he take only a few hours from his conference in order to attend the largest and most important relay carnival of the year. I condone his attending the conference and even recommended him as a candidate for Governor. I fully realise the value of attending such meetings. I do not, however, concur with the position that the Y.M.C.A. has placed Carl with their demands that he does not miss any time at their conference.

Carl's decision to attend the conference and miss the meet places in jeopardy the winning of the championship and shows no concern for the welfare of the many boys who also have sacrificed through the years in establishing and building themselves for an opportunity for a successful season. I recognise that the outcome of the meet may be entirely different than our hopes, but when our team competes in any meet, it is with the confidence that they are attempting to win.

It was clear I could not be Governor and run the Shorewood Relays. Madison was a good hour-long drive from Milwaukee, and I would need to be away from the conference for about six hours on the second day.

Coach had earned my respect, and many of my best friends were teammates. That said, our track team had real depth in its distance runners, and I was not a star runner. Serving as governor seemed most important. I informed Gehrmann of my decision. On March 14 he wrote to my parents,

Carl has been one of our finest leaders and has always shown concern for proper training habits and a willingness to sacrifice. Carl's predicament in this matter was not premeditated and he is a victim of circumstances. Therefore, because of my fondness for Carl and his welfare, coupled with my admiration and respect for the entire family, I shall reverse my original decision to drop Carl from the squad and invite him to join us with the understanding that I do not agree with the thoughtlessness of the Y.M.C.A. but do understand the predicament of Carl and his loyalty toward his promises to the delegates who elected him as Governor of the conference.

In response, my parents wrote,

We certainly appreciate your letter and realise how difficult it has been for you to make this decision. It is certainly the measure of the character of a man when he can change a decision which is for the benefit of another person. We regard you all the more highly for this unselfish act.

We, as well as Carl, are very happy that he will be able to participate in track this—his last—year in high school.

During lunch on Saturday at the model legislature, I was called from the head table to receive a phone call from my sister telling me that the Shorewood Relays were called off, to be run at a later date.

My final comment to close the model legislature:

As I look out into the assembly here, I wonder where each of you will be in ten years. What will be the condition of the U.S.? Will communism be a big threat, or will the threat have passed? Will democracy be on the rise or in down fall, fading from world picture? What will be the problems of the U.S.? Only time will answer. What we do will determine the answers. Failure of the majority to accept the challenge will mean disaster. If the youth of America, you and I, accept, our country will remain one of the greatest nations of the world.

2.7: Philleo Nash (Wisconsin Lieutenant Governor) and author.

Tosa, after dominating the indoor University of Wisconsin-Milwaukee High School Relays, won all of the meets in which it competed. We beat King 93.5 to 24; Kenosha and Waukesha 96 to 51.5 and 34.5, respectively; Greendale and Nicolet 118 to 53 and 41, respectively; Waukesha, West Alice Hale, and West Alice Central 73.5 to 61, 46, and 37.5, respectively; Whitefish Bay, Waukesha, and West Allis Central 97 to 65, 40, and 14, respectively. I ran the half-mile in all meets, always placing but never winning.

In the Suburban Conference Relays, with nine teams competing, Tosa not only won with 90.5 points but was also the first team to win all seven relay events, setting records in the low hurdle shuttle by 1.4 seconds and in the two-mile relay by 1.2 seconds in which I ran. I have no newspaper clipping from the Shorewood Relays; however, attached to my gavel from the YMCA model legislature is my second-place medal from the two-mile relay.

During my senior year, Gehrmann had two championship teams. I had become an accomplished runner and an integral part of those championship teams.

Chapter 3
Finding My Question

My writing the past two chapters and you reading them is extraordinary! Think for a few moments about what has just transpired. I chronicled with written words and documented with evidence events going back 75 years about which you now know. In the long arc of time, this has been possible for no more than one millionth of the time that life has existed on earth. Life appeared some 3.8 billion years ago, while modern humans evolved by several hundred thousand years BCE. By 10,000 years BCE, agriculture had permitted large enough aggregates of humans for written language to be created 5,000 years BCE, thereby enabling accurate histories to be recorded and known.

Thousands of years ago, peoples throughout the world told stories to explain whom they were, how they came to be, and why they exist. Every culture—many thousands of them—had a creation story that its members believed, giving the group coherence and identity. Could it have been different? The answer must be no. Although the human brain is essentially the same now as then, no one had an inkling of how the brain, senses, or anything else in the world worked. I came to Wesleyan to study science, the discipline that provided humanity with the methodology by which to establish human origins and the relationship of humans to the rest of life.

My first taste of science came in my senior year at Oberlin when I did a senior research project with Tom Scott, a plant physiologist. I learned about his research by writing a literature review on two classes of plant growth hormones—auxins and gibberellins—based on four review articles and 25 research papers that included Scott's first four publications. His concluding comment on my review was, "This does a nice job of bringing the ideas together. It is, however, written quite poorly. I have taken the liberty of hacking it to pieces so that you can get some idea of what I mean and what to expect when the time comes for thesis editing."

Auxin concentrations in the Alaska pea as affected by gibberellic acid, was the title of my research. The major objective was to measure diffusible and extractable auxin concentrations in pea plants both untreated and treated with gibberellic acid. Technical difficulties and time constraints allowed me to measure only short-term ether extractable auxin amounts and the heights of treated and untreated plants. I learned my first lesson in experimental biology—much needs to be learned before any useful data can be collected. In this instance, I needed to make growth chambers, learn how to grow pea plants, extract auxin, grow *Avena* coleoptiles, measure auxin concentration with the *Avena* curvature assay, and prepare reproducibly all the solutions required. Scott gave the project an A minus but, more important, I had been exposed to experimental biological research.

Scott knew Ian Sussex, a plant developmental biologist at Yale. After I had arrived at Wesleyan, Scott wrote to me to say he had written to Sussex, and that I should visit his lab. Yale's biology department had its department seminar on Wednesday afternoons, so I contacted Bill Frazier's father, whom I had met on my bicycle trip a year earlier. He invited me to have dinner with them after the seminar. Before the seminar I stopped by Sussex's lab to meet him and his graduate students. My Wednesday routine became biology seminar, playing handball and talking with Frazier followed by dinner at his home.

Donald Frazier was a 1933 graduate of Oberlin and minister of West Haven's congregational church. He had completed all but his dissertation in political science at Yale before shifting to a divinity degree so as to have better job opportunities. His intellect entertained everything from science to economics to religion, and we engaged in discussions across that breadth of human affairs. Are evolution and religion compatible or not? What does it mean to be a Christian? What is the meaning of life? Is investment in the stock market a good thing? Needless to say, I learned much and developed a deep respect for his assessments of the human condition.

3.1: The Reverend Donald Frazier.

I rented a room near Wesleyan for $40 a month from an Italian woman whose house was a short walk from Shanklin Laboratory. There I had a desk in the research facilities of Vincent Cochrane, chairman of Wesleyan's biology department. Cochrane was a noted microbiologist who wrote an acclaimed textbook, *The Physiology of the Fungi*. My Wesleyan fellowship was supported from his research funds.

When I arrived at Wesleyan, Cochrane was studying the enzymes of glycolysis, the biochemical pathway that breaks down glucose (a 6-carbon sugar) into two pyruvate molecules (each a 3-carbon sugar), yielding useful biological energy in the form of adenosine triphosphate. His enzyme studies employed the spores of the soil fungus, *Fusarium solani*, which causes disease in animals and plants. My tasks were to grow *Fusarium*, harvest spores, and then prepare from the spores an extract that contained the enzymes Cochrane was characterising.

I liked the job. I arrived around 6:30 a.m. to fix breakfast in the lunchroom adjacent to Cochrane's lab and have his enzyme extract ready for his experiments when he arrived at 9. It was similar to my two breakfast-headwaiter jobs at Oberlin, where I had started work at 6 and was off to class or to study by 9. By starting early, I had earned my keep before many students were even awake. By the late afternoon I was ready for a run, often a workout with Wesleyan's cross-country runners.

Holding a research fellowship instead of a teaching assistantship had its advantages. I had some control over when I worked, and, if I was efficient, the time required could be reduced. As soon as I had mastered growing the fungus and preparing enzyme extracts, Cochrane introduced me to the assays that he employed to study various glycolytic enzymes.

As the semester progressed, I found myself unexcited by enzymology and fungi. At night, as a study break, I would wander the building, talking with students and faculty. One night I walked by the basement office of a new faculty member, Spencer Berry. The room was dark except for the light from two fluorescent tubes in his desk lamp. On the arm of the fixture rested two moths, moving their large wings ever so slowly. I stood and watched, struck by their beauty. Berry acknowledged my presence and invited me in. I do not remember what we discussed other than the elegance of the newly emerged *Hyalophora cecropia* moths, but as I left the building to walk home, I knew these were the animals on which I wanted to do my master's thesis. Beauty won me over.

3.2: Professor Spencer Berry with a Hyalophora cecropia moth on his cigar.

What a rash thought! Cochrane, a senior faculty member and chair of the department, fully expected me to begin my thesis research with him spring semester. He was a seasoned mentor who had successfully guided the thesis work of graduate students. Here I was, abandoning his lab for that of a new PhD for whom I would be the first graduate student. Worried about upsetting Cochrane, I asked Scott for advice. He assured me that it likely would not be a problem. I talked with Berry about working in his lab and then steeled my courage to tell Cochrane. Scott was correct—Cochrane was professional and treated me with respect.

*****

Before arriving at Wesleyan, I had decided to run the Boston Marathon. To ensure that I would go through with it, I told several good friends my plan. The die was cast.

Boston is several hours from Middletown, and its marathon is the granddaddy of marathons, as well as the oldest continuously held long distance race in the U.S. Inspired by the Olympic marathon held in Greece in 1896, Boston Athletic Association team manager John Graham laid out the course the following year. Fifteen runners competed, with J.J. McDermott from New York who won the 1897 race in two hours, 55 minutes, 10 seconds.

I knew no one who had run a marathon, nor did I know how to train for a 26-mile run. My training log for the 1965 Boston Marathon began on February 20:

"Running condition fair to poor. Good for four miles in about 30 minutes with some pain afterwards. Recovering from a cold but began actual training work-outs today." I trained almost every day and by mid-March was up to eight miles at seven minutes per mile. By early April, I could run 12 miles at that pace. My total preparation entailed running about 200 miles over 55 days, with the longest distance being 12 miles.

I drove to Boston on April 19 in a mix of snow and rain. I stayed with a friend who lived near Boston. His family was delighted and excited to know their first marathon runner. I did not appreciate this adoration at first, but I would. The next morning, I headed to Prudential Plaza where the race would end, parked my VW, and took the bus to Hopkinton High School, where runners checked in and underwent a simple physical. The near freezing air chilled me as I checked into the prestart waiting pen. Like cattle, some 400 runners milled around, waiting to go to the starting line.

Bang! We are off. Nope—just a wise guy shooting off a firecracker. Then, bang for real! The herd slowly moved forward as the runners fanned out on the road. I had met a dozen or two fellow runners in the four hours before the start. As we recognised each other, we would run together and chat. As the first few miles slipped away, I joined two old pros—this was their fifth and sixth marathon—whose pace was to my liking. Soon they picked up several of their friends, and we had a real jolly group. With friendly talk and a joke now and again, we put the first 10 miles under in fine style: seven-minute miles. Soon Bob from the group and I eased away. A side ache I had had from the start subsided, as did a pain in the arch of my left foot. I was relaxed; the miles came easily. Could this last?

My left calf began to hurt somewhere between 11 and 12 miles. I was not too bothered; this had occurred many times before. At the halfway point, the friend who was hosting me in Boston handed me a bottle of dextrose that I had prepared as an energy drink. I chugged it down. Bob and I were going strong as we rolled off the 15th and 16th mile. Then a pain zipped up my left calf from ankle to knee. No, not now! Then again and again. I wished Bob the best as I slowed to a walk. My race was over.

Cramps! Good ole cramps. In high school, I had lain on the track, not moving until Coach Gehrmann brought rubbing alcohol and worked them out. I did not want this to be like that; I wanted to finish. I massaged my muscles, walked a bit, and resumed running. After a half a mile or so, the cramps returned. I repeated the process. Soon both legs cramped, but not so badly that I could not rub them out and run a little longer. And so it went for the last 10 miles.

It really was not so bad. Spectators on the side lines found my name in *The Boston Herald* next to my number (320) and urged me on by name. Little kids handed us oranges, candy, punch, and water; others passed out beer. It was truly a social event of the first order, not only for the runners, but also for the half-million people who lined the streets to cheer on Boston's runners.

I crossed the finish line exhausted and cold. Someone threw a warm blanket over my shoulders. I headed off to the showers and enjoyed a most satisfying bowl of hot beef stew. I was about the 150th runner to finish in about three hours and 30 minutes. I write 'about', because when I checked on my exact place and time, I discovered a break in the official record. Apparently, the cluster of runners finishing around 3.5 hours overwhelmed those recording the results. The last person listed in *The Boston Herald's* marathon results the next day was '143, Dave Vogel, Brooklyn, 3:27.13'. My pace for the first 16 miles was about seven minutes per mile, then 10 minutes per mile for the last 10.

The next morning, I could not walk downstairs from my second-floor room. I sat and eased myself down, one step at a time. That afternoon, I went to the gym; all I could do was walk around the outdoor track twice, a half-mile. I showered and went back to the lab.

I was delighted to have finished the Boston Marathon. "Everybody ought to run a marathon—once!" I told others. I continued running road races and ran three or four times a week to stay in shape and to relieve stress and solve problems. I had figured out early on that an hour's run was very effective therapy for a problem or frustration, as it gave my subconscious time to ruminate and often find a solution.

As the years passed, a voice inside my head started saying, "It is time to run another marathon." Five times I responded, and each time I learned more about marathon running. In 1984, I decided it was time to break three hours. I trained for six months and built up to a schedule the last month of running four times a week with three 10 mile runs, and one long run of 15 to 20 miles, for a month's total of about 200 miles.

Runners now knew about carbohydrate depletion loading. I tried it for my sixth marathon. Not only did the biochemistry make sense, it also worked! I am sure you have heard about marathon runners 'hitting the wall' somewhere around 20 miles—they will have run out of readily available energy.

Simply stated, here is what happens. In the first few miles, energy comes mostly from sugar in the blood. When blood sugar drops below 70 milligrams per 100 millilitres of blood, glycogen—a large molecule made up of glucose subunits stored in the liver and muscles—is mobilised as glucose into the blood by the hormone glucagon, from the pancreas. Around 20 miles into the race, sugar and glycogen are depleted in the runner's body; blood glucose level drops below that 70-milligram level. Glucose is scavenged throughout the body. Brain function decreases, as it is starved for energy. Simple math calculations become difficult, if not impossible. Muscles lacking glucose work inefficiently and build up lactic acid, which makes them stiff. Eventually they cramp. That is the wall I had smashed in to at the Boston Marathon.

The wall can be avoided, mostly by carbohydrate depletion-loading and sufficient distance training. Seven days before the marathon, I ran 20 miles to deplete glycogen from my liver and muscles. Then I went on a carbohydrate-free diet to prevent replenishment of glycogen in my liver and muscles; I could still eat meat and all the non-starchy vegetables I wanted. For each of the next three days, I ran 10 miles to remove any glycogen that had been synthesised and to keep my muscles starved for glycogen. All the while, my muscle cells responded by making more and more of the enzyme (glycogen synthetase) that produces glycogen from glucose. Because limited glucose led to essentially no glycogen, the glycogen-synthesising enzyme continued to pile up way beyond the normal level—the enzyme was what biologists call 'super induced'. This metabolic trick worked because no newly synthesised glycogen was present to provide negative feedback on the gene for glycogen synthetase, thereby turning off the synthesis of glycogen synthesising enzyme. Running under these conditions is horrific, similar to how a marathoner feels in the last miles. Meanwhile, my body was internally haemorrhaging. The stretch sensors from a full stomach inform the brain, "You're full. Stop eating!" while blood glucose sensors tell the brain, "You're starving. Eat!"

On the fourth, fifth, and sixth days after the 20-mile run, I limited my work out to a warm-up only. On the fourth day, I began loading and made two pecan pies, one for my lab group and the other for me. I continued eating plenty of carbohydrates through the sixth day. The surge of carbohydrates was metabolised into glycogen by the excess of glycogen synthetase; so as to super charge the muscles with energy.

I ran the 1984 Hudson Mohawk River Marathon, my seventh and last marathon, the next day in 2 hours, 55 minutes, 20 seconds, a pace of 6:43 per mile, essentially the winning time for the first Boston Marathon 87 years earlier. When I finished, I was so expended that all I could do was cry. Our daughter, Virginia, who had watched me run other marathons, asked my wife Mary, "What is wrong! Why is Dad crying?"

Mary responded, "Nothing. He is just really happy." And I was. I no longer run, but I have deep memories and pictures.

3.3: Author running Hudson-Mohawk Marathon at mile 18 and finishing.

*****

Enough of marathon science. Back at Wesleyan, Spencer Berry's research focused primarily on the biochemistry of the development of silk moths. He studied the synthesis of ribonucleic acid (RNA) and deoxyribonucleic acid (DNA), molecules involved in protein synthesis and genetic information. My interest was in whole animal developmental physiology.

Insect development takes place in a step-wise manner that is imposed by the basic insect-body structure. An animal with an external skeletal structure can grow only until it has filled its exoskeleton. Before further growth can occur, the animal must moult, a process by which a new, larger exoskeleton is laid down inside the old exoskeleton, which is then shed in the moulting process. The number of moults from egg to adult varies from species to species, and patterns of differentiation are of two types. In one case, the larval stages are morphologically similar to the adult until the last moult leads to a sexually mature insect, crickets are an example. In the other, the larval stages are morphologically distinct from the adult until the process of metamorphosis leads to a sexually mature adult. The saturniid silk moths that Berry studied (*H. cecropia* and *Antheraea polyphemus*) exhibit the second pattern, undergoing four larva-to-larva moults followed by a larva-to-pupa moult and then a pupa-to-adult moult that produces the fertile insect. The pupa of these moths, prior to the final moult, enters into a developmental arrest period called diapause, which is common among insects. Diapausing insects have extremely low metabolic activity. Diapause in *H. cecropia* and *A. polyphemus* enables these species to survive winter conditions and permits a coordinated emergence of short-lived fertile adults in spring, thereby facilitating mating at a time when larval food is available. I was interested in the hormonal control of the last moult, which completes metamorphosis.

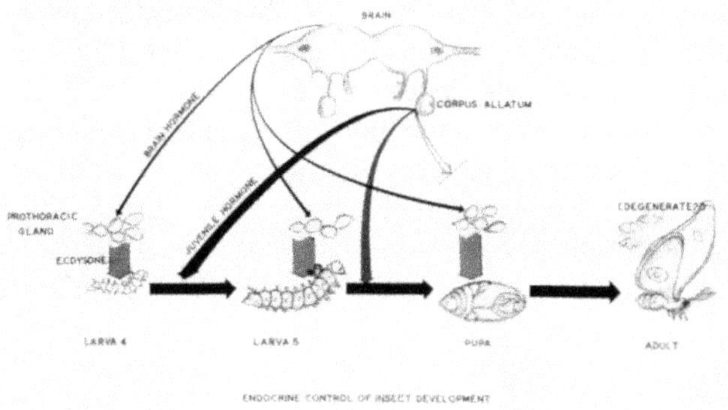

3.4: Diagram of life cycle of Hyalophora cecropia and its hormonal control.

Moulting in insects that undergo metamorphosis is controlled by three hormones: brain hormone produced by hormone secreting cells in the brain, juvenile hormone produced by corpora allata (two glands each connected to one of the two halves of the brain), and ecdysone, made by the prothoracic glands. Brain hormone activates the prothoracic glands to produce ecdysone, while the concentration of juvenile hormone determines the type of moult. High juvenile hormone concentration leads to a larval–larval moult while a low-level results in a larval–pupal moult. An absence of juvenile hormone results a pupal–adult moult.

My master's thesis research followed up on two observations. First, the implantation of chilled-brain-neurosecretory-cells into diapausing *H. cecropia* terminated diapause. Second, the photoperiod, or the hours of light and dark in a 24-hour period, was a major factor in the termination of diapause in *Antheraea pernyi*, another saturniid moth closely related to *H. cecropia* and *A. polyphemus*. My work examined the effects of photoperiod and brain implants and transplants on diapause termination in *A. polyphemus.* I would also attempt a concurrent histological study of brain and eye development in intact individuals and in animals with implanted and transplanted brains. By doing a histological project, I would have something to present in my thesis if the experimental work yielded ambiguous results.

With input from Berry, I designed and performed on *A. polyphemus* 20 experiments along with correlated histological analyses. I summarised my 63-page thesis with six conclusions from the experimental work and five from the histological study. The histological work was instructive and added to the thesis, but it was not sufficiently new for publication, whereas the experimental part was of high quality. Over the next year, Berry, with my input, prepared a manuscript on the experimental studies that was published by *Nature*, an international journal of the first rank. The *Nature* article concluded with:

These experiments confirm that the brain can be activated in *polyphemus* by surgical procedures which do not involve chilling the animal. It is also evident that the prothoracic glands can be activated directly by injury and low temperature in the absence of a brain. The fundamental picture of the brain activating the prothoracic glands remains intact, but in exceptional circumstances other agents may be effective.

What we did not recognise at the time was that these experiments gave some of the first data indicating that ecdysone acted on the prothoracic glands to promote more ecdysone release, or an instance of positive feedback. Today this ecdysone positive feedback in metamorphosis of insects and that of thyroxin in metamorphosis of amphibians are textbook examples of positive feedback.

*****

A young black woman going house to house came by our rooming house in the fall of 1965, telling whoever answered the door about the tragedy of the Vietnam War. She absolutely believed that the U.S. was making a monumental blunder and had no business being there. I listened, but had little to say, as I passively trusted the government. I now know she was far more informed than I was.

I had become tired of being a student. The Milwaukee draft board had me classified 2E through June 1966, meaning I would be drafted as soon as I finished at Wesleyan, unless I stayed in school. My freshman and senior roommate at Oberlin and good friend, Bill Fogerty, was enthralled by the story of *The Caine Mutiny*, the 1951 Pulitzer Prize-winning novel of Herman Wouk on the fabricated story of *USS Caine* and its incompetent captain. He was bored with teaching junior high school students and thought that we should become naval officers. The *Caine's* story did not motivate me, but joining the navy made sense to me because of my middle name, in addition to my brother already being in Vietnam as a swift boat captain. I deeply believed that many had gone to war for the freedom I now enjoyed, and I accepted that it was my responsibility to do the same.

Bill Frazier graduated from Oberlin in the spring of 1965 and married Mary Thurston, a woman he met when at Oberlin, in June. His sister, Mary, was a bridesmaid, and I a groomsman. Over the past year, during my Wednesday visits, I had gotten to know and liked the Frazier family, including Mary. After the wedding, she and I chased the newlyweds in my VW bug, beeping the horn—even driving on the sidewalk to stay behind them. That evening, Mary and I saw *The Sound of Music* together, our first date.

Mary's mother commented more than once that Mary gave the boys she dated endurance tests. Could he bicycle 50 miles on Long Island? Climb Sleeping Giant or East Rock near West Haven? Swim half an hour in Long Island Sound? Apparently, I turned the tables on her. We did all of those things, and more. One Sunday morning we walked a total of 14 miles to and from her house to the top of East Rock before church. She wore high heel shoes, and her feet hurt so badly that she had to finish the walk shoeless. Her injured big toenails eventually fell off.

I was raising *A. polyphemus* moths under netted maple trees on a farm near Middletown to have pupae for my fall experiments. Sometimes Mary came with me to check them. It was there on a beautiful summer day that we first kissed in a field of wildflowers and summer grasses.

Although Mary was a sophomore at Cedar Crest College in Allentown, Pennsylvania—200 miles away—we saw each other whenever we could. I taught her how to drive stick shift so she could drive the bug. We walked and picnicked on the Appalachian Trail near Cedar Crest and explored Indian Ladder near West Haven and Wadsworth Falls in Middletown. One Saturday morning, we hiked up Sleeping Giant so early and fixed breakfast in the park that the ranger accused us of being there all night. We saw *The Fantastics* at Southington playhouse summer theatre; heard Peter, Paul, and Mary in concert at Yale Bowl; watched movies—*My Fair Lady*, *Westside Story*, and *How the West Was Won*; and had breakfast and dinner at our favourite restaurants. We were falling in love!

3.5: Mary sitting on top of author's VW bug packed with all author's possessions.

After defending my thesis, I packed everything I owned into my bug, tied a trunk to the top, and used a bike rack I had designed to attach my bicycle above the rear-engine hood. With just enough space left in the front seat for Mary, we drove to Wisconsin so she could meet my family. In July, we drove to Newport, Rhode Island, where I had orders to report to U.S. Naval Officer Candidate

School (OCS). Mary drove the bug back to West Haven. She visited me in Newport a few times before returning to Cedar Crest.

OCS was established in 1951 to fill the need for junior officers to man the World War II mothballed fleet that was restored at the outbreak of the Korean War. The four-month curriculum included programs of general orientation, engineering, seamanship, navigation, and naval justice. When I attended OCS, it was supplying more than 50 percent of the Navy's junior officer ranks, which were swelling with escalation of the Vietnam War. Four of every 10 applicants to OCS were selected, leading to commissioning nine out of every 10 candidates. At the age of 24, I was about the average age of the 345 successful candidates in class 701. Essentially all were college graduates, many with advanced degrees. The Vietnam draft had led to a very talented cadre of new naval officers.

I was among 24 candidates assigned to Juliett Company of the Third Battalion berthed in one of the original barracks. Just prior to our arrival, the 940-man Nimitz Hall dormitory with two-person rooms opened as the first new barracks since OCS commissioning. I was glad to be in the old barracks, where we lived in a large, open room with upperclassmen of Juliett Company, because it was radically different from college dorms and allowed for 24-hour bonding. From the moment reveille sounded at 5:30 a.m. until taps at 10:30 p.m., every minute was programmed, and 701 Julietts did it all together. We marched as a group and were expected always to exhibit military bearing. New candidates were required to request permission from an upperclassman for almost everything, even a bathroom break. If we did something inappropriate or poorly, an upperclassman might reprimand us, give demerits, or ask for something physical, such as 25 push ups. We were restricted to the naval base for the first four weeks but were given free time Friday evenings from 5:00 to 8:30.

I adjusted easily to the rhythm and restrictions of OCS. In truth, it posed the same challenges I had faced in every new situation. What are the rules? How do I fit in? However, it was the first time I came to appreciate the true meaning of freedom. To imagine living this way permanently was a dismal prospect.

After our plebe month, we had off-base liberty from 11:00 a.m. Saturday to 7:00 p.m. Sunday. Mary visited me before returning to college for fall semester; her visits were a most welcome change. It was not that I did not like OCS or found it difficult; in fact, I thrived. I did well in class and was in excellent physical condition, although I did not much like jumping off the 12-foot-high platform into the pool to simulate abandoning ship.

We had lots of hands-on experiences. The most relevant of these experiences was the tour of a destroyer, especially the boiler and engine rooms. As I climbed the ladder to sunshine, I decided that I did not want engineering duty on a destroyer, or any other ship, having seen the cramped spaces filled with pipes and wires going every which way along with myriad valves and pipes connecting numerous types of equipment.

My orders came in late October: "… If found physically qualified and upon acceptance appointment, proceed port in which USS RICHARD B. ANDERSON (DD-786) may be, arrival report CO, duty." My brother and others told me that a destroyer assignment was coveted duty for a newly commissioned line officer. *Anderson* returned to San Diego on December 16 after major repairs in the Long Beach Naval Shipyard. The next day I reported for duty. I was informed that I would be electrical officer in the engineering department until I completed three months of engineering school, after which I would be the Main Propulsion Assistant (MPA) in charge of engine and fire rooms. In retrospect, this assignment was perfect for me. I was busy doing concrete things and knew when I had done my job—the ship got underway, the galley had water, lights were on, and showers had hot, fresh water.

3.6: Author in dress whites on USS Richard B. Anderson DD786 in Tonkin Gulf at the time of a change of command ceremony.

Mary flew stand-by to San Diego three times that winter and spring. At the downtown Congregational Church on Easter, the greeter, Marguerite Miller, welcomed us as strangers. As long-time residents in this navy town, Marguerite and her husband, Dick, had welcomed many sailors. They invited Mary to stay in their downstairs apartment. It was a wonderful arrangement that perhaps influenced us, as we have had many dozens of people stay in our home during our 50-plus years of marriage. We became good friends with the Millers, visiting them when in San Diego and exchanging letters until they passed away.

I started engineering school at the end of March and moved into the San Diego Naval Station bachelor officer quarters. *Anderson* deployed for West Pac in late April; I would join her in June after completing engineering training. In June, Mary came for her third visit. We spent all my free time getting to know San Diego and hiking in Cuyamaca Rancho State Park and Anza Borrego Desert.

On our last hike at Cuyamaca, we sat and talked in a field of wildflowers. I made a ring of flowers. Proposing, I put it on her finger. With a kiss, she accepted. Following a tradition now long past, I had written to Mary's father and asked his permission to marry his daughter. Her parents' approval arrived in a letter a few weeks before our final hike in Cuyamaca.

On the last day of her visit, I drove her to the airport and then to a Volkswagen dealership. They had agreed to purchase my bug for $450, a good deal, I thought, as I had paid $800 two years and 50,000 miles earlier. They even commented on how good the tires my dad had purchased looked. I mailed the check to San Diego Savings Bank before heading to the airport for my flight to Clark Air Force Base on Luzon Island, Philippines.

Over the next four months, *Anderson* provided shore bombardment off the coast of Vietnam in support of our troops and to disrupt the ability of Vietcong and North Vietnamese to move supplies up and down the Ho Chi Min Trail. We worked with attack carriers, providing a reference point for incoming planes by positioning *Anderson* 1,000 feet astern and slightly to the portside of the carrier. This way a pilot would know if he was at the appropriate speed and altitude for a safe landing. If a plane had to ditch on take-off or landing, we were ready—although that never happened while I was on board. *Anderson* also spent time in the Tonkin Gulf off the coast of North Vietnam, where we monitored incoming and outgoing carrier planes that bombed Hanoi Harbour or other targets. When a pilot was hit, his goal was to 'get his feet wet', so that a destroyer could pick him up before a North Vietnamese boat did—a pilot was a prized acquisition. We rescued two pilots during my time on *Anderson*. If a pilot had to eject over land, he needed to be rescued before being captured. One time we refuelled a rescue helicopter, only for it to be shot down and the pilot captured. At times the North Vietnamese would use a downed pilot as bait for a rescue helicopter, so they not only had the pilot but also a helicopter crew.

Anderson had a complement of around 300 men who lived in close quarters when at sea. We were a modest sized community squeezed into small spaces that provided living space, food, recreation, water, electricity, entertainment, basic health care, and a rule-based safe environment. Each person fulfilled a basic function and contributed to others, mostly on a volunteer basis. I was MPA as well as chaplain and athletic director. It did not take long for me to know the 60 men in my division and many others, because we mustered every morning to ensure that no one had gone overboard or was otherwise absent. The crew sized-up each officer and gave him a nickname. Mine was Forty Knot. I took this positively because the top speed on a destroyer of our class was 35 knots.

The ship's routine while underway had two daily patterns. When in the combat zone, two watch groups stood four hours on, four hours off. At other times, three watch groups stood four hours on, eight hours off. In both cases, ship's work began at 8 a.m. and ended at 4 p.m. After three days, our watch schedules shifted by four hours—the 8 a.m. to 12 p.m. watch group shifted to the 12 p.m. to 4 p.m. time period. This shifting of watch times was terrible for me because my biological clock was slow to reset, and I would get sleep deprived.

Like other officers, I often had to choose food or sleep. A noon nap in lieu of lunch was often the choice.

Needless to say, we got to know each other well. We depended on each person doing his job well; otherwise, things could go drastically wrong, as we saw in 2016 when both *USS Fitzgerald* and *USS John S. McCain* collided with commercial vessels. Every sailor and officer can cite at least one case with a thin line between disaster and being OK.

For me this came one night, *Anderson* was steaming with *USS Warden* in the Tonkin Gulf under the condition of 'darkened ship'—no running lights or other visible lights. It was the midnight to 4 a.m. watch. I was the Junior Officer of the Deck (JOOD) and had the Conn, meaning I instructed the helmsman as to direction and rudder position, and the engine-order-telegraph-operator the direction and speed of each propeller. A more senior officer on the bridge was the Officer of the Deck (OOD), who instructed the JOOD and could take the Conn, if needed. All of a sudden, Combat Information Centre informed the bridge that *Warden* was constant bearing, decreasing range. We were on a collision course with her! With my binoculars I saw a dark shadow off the starboard bow. Without consciously thinking, I ordered, "right full rudder, port engine ahead full, starboard engine emergency reverse," and said loudly into the voice tube to the captain's sleeping cabin right off the bridge, "Captain to the bridge." By the time he came to the bridge—no more than a minute later—*Warden* was passing across our bow at perhaps a hundred yards. The captain, seeing that the crisis had passed, said calmly, as if nothing had happened, "Resume station," and left the bridge. Of course, he would want a full report in the morning. Under 'darkened ship' conditions, speed and course changes are signalled by Nancy Light (an infrared signal light that cannot be seen at night without special equipment). Our signalmen had missed a course change from *Warden*, our senior ship from whom we received orders. Had we collided with *Warden*, many would have died or been injured. As a result, the lives of the captain, the OOD, and myself would have been very different if we would have survived.

Mistakes, as well as good decisions, have consequences. On August 7, 1964, the U.S. Congress had passed the Tonkin Gulf Resolution, which committed our country to the Vietnam War based on 'phantom' attacks on two destroyers. It was the reason for my presence in the Tonkin Gulf, and worth revisiting.

In the mid-afternoon of August 3, 1964, *USS Maddox*, operating in the Tonkin Gulf off the coast of North Vietnam, had been involved in an attack by three North Vietnamese torpedo boats provoked by actions of the South Vietnamese Navy over the previous days. According to standing orders, the *Maddox* engaged the boats, as did four F8 Crusader aircraft from *USS Ticonderoga* called in by *Maddox*. Two torpedo boats were heavily damaged while the third was left burning and dead in the water. Because the South Vietnamese Navy caused this incident, it was not considered an unprovoked attack on the *Maddox* and thereby on the United States.

The next day, August 4, U.S. intelligence obtained evidence that the North Vietnamese Navy intended to conduct offensive marine operations in the Gulf of Tonkin; the exact details were unknown. Thunderstorms, squalls, and six-foot seas reduced visibility. That evening, the *Maddox* and a companion destroyer moved 100 miles off the coast to gain manoeuvring room in case of an attack. At 10:40 p.m. the *Maddox* reported radar surface contacts that appeared to come at the ships from the northeast, southwest, and east, behaving like attacking torpedo boats. Targets disappeared, then new ones appeared. Over three hours, the ships maneuvered to evade apparent enemy attacks. During this time, the destroyers fired 249 5-inch shells, 123 3-inch shells, and dropped four depth charges. When the *Maddox* engaged the apparent enemy boats, Captain Herrick, commander of Destroyer Division 192, was on the ship and in charge. He reported the apparent attack to his command in Honolulu. Before the engagement ended, however, Herrick had doubts about the authenticity of the radar contacts and the sonar detections of torpedoes.

After the attack, Herrick reviewed the events with *Maddox* crew and sent the following message to Honolulu that Washington received at 1:37 p.m. on August 4, 1964. "Review of action makes many reported contacts and torpedoes fired appear to be doubtful. Freak weather effect on radar and overeager sonar men may have accounted for many reports. No actual visual sightings by MADDOX. Suggest complete evaluation before any further action taken."

Commander Stockdale confirmed this conclusion. He was one of the four F8 Crusader pilots who had engaged the three torpedo boats a day earlier and who had observed the last 90 minutes of the purported engagements with the *Maddox.* He reported later: "I had the best seat in the house to watch that event, and our destroyers were just shooting at phantom targets—there were no PT boats there… there was nothing there but black water and American firepower."

President Lyndon Johnson and Secretary of Defence, Robert McNamara, did not take the time to carefully and thoroughly assess what had transpired on August 4. The next day they ordered a retaliatory attack on North Vietnam, beginning years of conflict. Commander Stockdale on the *Ticonderoga* was ordered to lead the strike force of 18 aircraft on that first attack, knowing "we were about to launch a war under false pretences, in the face of the on-scene military commander's advice to the contrary."

On August 7, Congress passed the Tonkin Gulf Resolution, authorising the president to "take all necessary measures to repel any armed attack against the forces of the United States and to prevent further aggression." Congress had abdicated its Constitutional role of approval and oversight of U.S. military forces. President Johnson summed up what had happened: the Tonkin Gulf Resolution "was like Grandma's nightshirt. It covers everything."[2]

*****

---

[2] https://www.usni.org/magazines/navalhistory/2008-02/truth-about-tonkin. Accessed, January 28, 2018)

Anderson returned to San Diego at the end of October. I took five days leave in November to fly east and visit Mary and review the wedding plans she and our mothers were making for our December 22 ceremony. During the month of December, *Anderson* was ready-ship at San Diego Naval Station. It had to be underway in six hours if called. The ship's crew was available with boilers powering the ship.

On December 5, *Anderson* was ordered underway. Working in engineering, I did not have time to call Mary. It did not matter, however, as I did not know where we were going or for how long. The mission turned out to be shadowing and jamming the communications of a Russian intelligence-gathering vessel that was within 200 miles of the U.S. coast. For a week, we observed them from several hundred yards until we had to break off for refuelling. We reached Astoria, Oregon, on December 15 at around 2 a.m. The bunker-C fuel oil available was so viscous that it took more than three hours to pump; refuelling at sea was much faster. I broke free to call Mary at a little after 5 a.m. I told her the ship had been mobilised and I did not know when I would get back to San Diego. If she did not hear from me by December 20, she should postpone the wedding. At church that night, her mother was asked about the wedding, to which she responded, "The wedding may be postponed. The groom is somewhere at sea." Fortunately, the Russia ship moved outside the 200-mile limit the next day and we were relieved by aircraft, which continued surveillance. *Anderson* docked at San Diego in time for me to fly back east on December 20.

3.7: Wedding picture of Mary Frazier McDaniel and author.

Before the wedding, Mary's father took us into his office to offer financial advice. He encouraged us to save up until we had at least six months of living expenses set aside. He went on to explain that most people are held hostage by their jobs; six months of living expenses in the bank allows you the freedom to refuse something if it violates your ethics or principles. Anybody can get a job in half a year. Mary and I took his advice.

My dad, in fact, had given me similar advice when I was in high school, based on his Depression-era experience. "No matter what you make, save at least a nickel of every dollar. The real goal should be 20 cents, but at times that might not be possible." When I was 6, in 1948, my dad introduced me to savings by opening an account for me with a deposit of $9. By 1956, when our family moved to Wisconsin, I had saved $213.37—reflecting small deposits I had made from money I had earned or was given. In high school I worked in the student cafeteria and at the snack bar after athletic events. These and other savings amounted to almost $1,000, which paid for my first semester at Oberlin. For my other college expenses, my dad said, "You earn as much as you can. Mother and I will do our best to cover the rest."

*****

I had no idea how difficult the academics at Oberlin would be. At first-year-orientation we were told, "Look to your left. Look to your right. In four years one of you will not be here." That was foreboding! I decided I had better not look for jobs until I knew more. As the first several weeks passed, I was not overwhelmed; rather, it felt more like summer camp and other new situations that I had managed by working hard. I took a job in my dorm answering the phone several hours a week. Then the headwaiter at Dascomb, where I ate, asked for students who might want to work as substitutes when a regular crew member was sick. I volunteered. Soon I was asked to fill in regularly. During Thanksgiving break, I joined a few others to strip and wax the dining room floor. At the end of the semester, Kit Salter, a senior who worked with our dining services provider, asked if I would be interested in a breakfast headwaiter position. "Yes!" I said. Kit looked me in the eye, advising, "There are only half a dozen on the crew, and most are juniors and seniors. You are a freshman. For a month, just do your job. Don't tell any of them what to do; they know their jobs. Do just what I have seen you do: show up early, stay until everything is done, do your job, and work hard with a smile. You will gain their respect, after which you can make changes." Kit gave excellent advice and I did my best to follow.

During sophomore year I was breakfast headwaiter at a larger facility, then a lunch-dinner headwaiter during my junior year. A women's dormitory opened my senior year, where I was lunch-dinner headwaiter. That required me to write job descriptions and manage waiters and waitresses, dishwashers, and a clean-up crew—totalling more than 50 students. I earned full board all four years, but when in charge of lunch-dinner, I earned an extra $105 and $200, respectively.

My cost for four years at Oberlin was $8,715 for tuition, room, board, and fees. All other expenses, including books, transportation, clothes, entertainment, and snacks came to $1,325—for a total of $10,040 (I kept a log of every expense over my four years at Oberlin). I had been able to earn just over $5,000; my parents covered the rest.

At Wesleyan, my stipends while earning my master's degree totalled $4,000, of which I saved $800. I was so focused on living within my means and saving money that I ate hot cereal for breakfast. I had calculated that hot oatmeal cost 4 cents per meal as opposed to 5 cents for cold cereal. During the summer, I took care of a fraternity house for a free room and worked as a lifeguard at the college pool. I used my savings to purchase my first stock, 10 shares of Texaco—I had yet to know that burning fossil fuels would foster global warming.

*****

I sold my bug while in San Diego, not only because I knew that Mary's parents worried about her driving such a small car on the California freeways, but also because they would be giving us a Chevy Impala station wagon as a wedding present. It was a big car. We named it The Pig because of its 16 mpg—my VW bug delivered 40 mpg. However, it was well made, ran smoothly at 100 mph, and had gone over 100,000 miles when a road-grader caused an accident that totalled it. Stuart, our 3-year-old son, was in his car seat when it happened. After sliding into a tree that launched the car into a field, Stuart said to Mary, "Tell that naughty road-grader to get out of the way."

After our wedding, Mary and I drove to Wisconsin to spend Christmas with my family and then on to California. Mary returned to Cedar Crest in January to complete her degree. Many navy wives were teachers, as was Mary. She secured a position teaching first grade at Rohr School, just north of the Mexican border. The families were poor, and many parents did not speak English. When parents did not come to school for conferences, Mary worked with a colleague who taught English-as-a-Second-Language to find a time during a weekend to meet at the school with the parents. During times when *Anderson* was in port, Mary dropped me off in the mornings and picked me up after school. It was a stressful start to teaching. We purchased a used piano for her to play so she could relax after a day at Rohr School.

Inspired by a fellow officer and his wife, who had hosted weekend breakfast gatherings for the wardroom (the ship's officers) when I was a bachelor, Mary and I began a practice of entertaining over breakfast. We liked breakfasts because alcohol was not expected, and the food options were inexpensive and easy to fix. When I was promoted to lieutenant junior grade, Mary and I hosted an atypical 'wetting-down' party for the wardroom—we served homemade ice cream instead of alcohol. *Anderson*'s officers never gave me a hard time about not drinking and enjoyed the ice cream. I had kept the promise I had made to myself when I was 10, but that was about to change.

In the Navy, drinking and smoking were inexpensive social habits. When *Anderson* entered international waters, a carton of cigarettes cost just one dollar, while drinks—whisky sours, black Russians, martinis—were 10 cents during happy hour at the officer's club. And while I know smoking does have positive short-term effects—increases alertness, sharpens short term memory, gives mild euphoria and feelings of relaxation—the long-term effects are costly and many: an increased risk of cancer and stroke, chronic bronchitis, emphysema, shortness of breath, and heart disease. The most seductive aspect of these activities may be the positive social interactions that these activities engender.

In June 1968, midshipmen from the Naval Academy boarded *Anderson* for their summer cruise. After training exercises, our port of call was Pearl Harbour, my first visit to Hawaii. I was now among the senior officers in the wardroom with whom I had bonded. We were comrades. Renting a VW bug, several of us toured the island, stopping for walks on the beach or swimming. In the evenings, all except those standing watch went to the Officers Club for happy hour. The social aspect of drinking soon had me joining the dice rolls to see who would buy the next round of drinks. My frugal nature was not offended. If there were seven in the dice roll, the winner had a mere 70 cents to cover.

After Hawaii, we continued to expose the midshipmen to various activities, including operating with other ships and required tests. I was bosun-chaired to the *USS Bausell* as the official observer for a full power test. On completion, the *Bausell* asked to keep me, as both ships were headed to San Diego. Our communications officer replied with a visual message: "UNCLAS YOUR 192323Z AFFIRM KEEP MPA UNTIL ARR PORT. HE IS ALSO A QUALIFIED BIOLOGIST IF YOU HAVE ANY BUGS REGARDS DOUG."

On watch we passed slack time with idle stories. I talked mostly about the animals we would see on watch (porpoises and flying fish), single-celled organisms that luminesce when disturbed by the ship's wake, or the metamorphosis of a caterpillar into a butterfly. The ship harboured a healthy population of cockroaches. The stewards never left food uncovered in the wardroom or its galley because within minutes cockroaches would be feasting. I would engage roach detractors with information on roach behaviour; after all, every organism has the same challenge—survive long enough in an unfriendly environment to produce more of its kind. I took Doug's comment as an affirmation of bringing biology onto *Anderson*'s bridge.

As summer drew to a close, we completed preparations for deployment. Our wives, soon to be West Pac widows, gathered at Point Loma. It overlooked the channel the ship passed through to reach the Pacific. As we passed them, I had the firemen signal farewell with a puff of smoke.

*****

In the Navy Anderson took me to Hawaii, Hong Kong, Japan, and the Philippines. Experiencing people and their cultures in these places gave depth to my understanding of diversity that is difficult, if not impossible, to gain from reading or images alone. Yes, my exposure was brief and superficial. Nevertheless, it was a beginning on which I would build. The people and culture that I did experience in much greater depth were the vast majority of Americans who were well represented in the 60 enlisted men—firemen, mechanics, and electricians—in my division. I was responsible for their professional development, mediated mostly by higher ranked enlisted men and the two navy chiefs in my division. I met with all of them casually as they worked but also formally for required evaluations and when problems arose. Their values, aspirations, and to some degree, ethics, were often different from mine.

Many had no capacity to save money. Like today's payday loan businesses, the enlisted men had one of their own—under the table. Payday came every two weeks, and most enlisted men took their wages in cash. The rate for a loan was $5 today for $7 next payday. Enforcers, who were close at hand on paydays, kept the system running smoothly. With a rate of 40 percent per two weeks, it was easy to get into unmanageable debt. Young sailors were also hustled by barmaids. I had a sailor who 'fell in love' and purchased an engagement ring for which he had to borrow cash just before we deployed. A letter soon came from the woman breaking off the engagement, but with no offer to return the ring.

In Kaohsiung, Taiwan, a sailor in my division failed to return to ship one night. Shore patrol picked him up the next day, and he was restricted to the ship. That night he jumped overboard and swam to shore to find 'his' girl. Shore patrol brought him back, this time to the Marine Brig on the tender in the harbour. He was returned to *Anderson* after a day because we were getting underway. He had a marine haircut (shaved head) and a much-improved attitude. Most people in the military know that being absent without leave (AWOL) is a serious violation, but some test the system when judgment goes haywire. These situations were indicative of the differences in values held by some of the men in my division and me. Even though I had gone to a big urban high school with almost 2,000 students, it was my experience as a division officer that truly awakened me to just how different I was from the majority of Americans in the 1960s. And now, after a career as a scientist and educator at a research university where the separation from the ordinary person on the street is extreme, I have even less in common with most Americans.

*****

Knowing my tour of duty on *Anderson* would end in December, I sought reassignment to a billet that would take advantage of my master's degree in biology. In my request to the Bureau of Naval Personnel (BuPers), I wrote on August 6, 1968:

Before entering the Navy, I earned a master's degree in Biology, and since have published some of the results of my research in *Nature*. My interests are in science and I plan to enter a Ph. D. program in Biology when my tour of duty with the Navy is completed. It is my belief that I would perform in a more outstanding manner if my assignment was more scientific in nature, and I know

I would be more satisfied personally if I was putting to explicit use my six years of higher education.

In October I received a postcard from BuPers stating that I would be assigned to the U.S. Naval Academy as a chemistry instructor. I was detached from *Anderson* in early December, in time for us to drive to Wisconsin to spend Christmas with my family and then New Years in Connecticut with Mary's family. I reported for duty in the science department on January 3, 1969. The day we arrived, Mary secured a position teaching pre-first grade in Severna Park, Maryland. We rented an apartment within walking distance from the Academy, and I started preparing to teach introductory chemistry. Soon after, Quarters M at the naval station became available, and we had our first house. I dug up a patch of grass on the south side of the garage; Mary and I had a garden—the first of 49 over 50 years.

The chemistry department had a faculty of 35, half civilians and half officers. I taught three sessions of introductory chemistry to about 30 midshipmen each. We had a common textbook and syllabus of topics, but we individually prepared presentations and tests. In my college experience, chemists and physicists appeared to give much harder tests than biologists. I was curious about the relationship between a test's difficulty and student performance. After the first test, I decided to make the second test much harder and the third test easier to see if the relative ranking of a student changed.

As expected, students scored significantly higher on the easy test. Students near the top or bottom of the class retained their position in both tests, while students in the middle showed a little less consistency. Believing that students felt better with a high-test score and were not necessarily motivated to work harder by a low score, I leaned toward easier tests then and thereafter. I still believe that was a good decision, but my evidence is slim.

I really enjoyed teaching my own classes. Mary, too, was having a positive experience. We quickly settled into Annapolis life. We joined the local United Church of Christ, I routinely ran road races, and we socialised with other officer families living at the Naval Station, often by catching crabs in the Severn River and eating together at one of our homes. I was accepted to the biology PhD program at the University of Maryland, with an interest in researching erythropoietin, the hormone involved in mammalian red blood cell production.

I had been playing handball since my college days, including matches with Mary's dad on days I would attend seminars at Yale. It was an enjoyable physical and social activity. I began playing regularly with another officer in the science department, soon after arriving at the academy. He and I were evenly matched often splitting the first two games, but he was a heavy smoker. At first, I often won our set of three games, because he would tire in the third. In 1965, the U.S. Surgeon General began publishing *Smoking and Health*, a set of reports that linked smoking to lung cancer. At the time, the federal government was airing a major antismoking campaign on TV to influence a child's attitude toward smoking. In one ad, after explaining the ills of smoking, the child would ask, "Daddy, why do you smoke? Don't you love me?" In late spring, I began losing more frequently. "Why have you gotten so much better?" I asked. He replied, "I just got tired of my children asking me why I didn't love them every time I lit up a cigarette. I quit smoking." Forty-two percent of adults in 1964 smoked; in 2015 the percentage had dropped to 15. Piles of data now affirm what my fellow officer experienced—properly designed messaging over time can significantly reduce detrimental behaviours like smoking.

During the 1968-69 academic year, the science department initiated a pilot biology program, with one section of 12 midshipmen taught by a civilian instructor who held a master's degree in biology. Enrolment in the fall 1969 biology program totalled 165 midshipmen who filled seven biology sections, including a laboratory. As the only member of the science department with an advanced degree in biology, I was delighted to shift from chemistry to biology. In a request dated September 5, 1969, to the chief of Naval Personnel via the superintendent of the Naval Academy, I asked for a one-year extension of active duty from August 1970 to August 1971. My justification was:

At the present time... I am the only officer [in the Science Department] having a Biology Background. [A civilian instructor and I] are working together to develop and instruct this expanding program. I believe that it would substantially benefit the Biology offering at the Academy if I were able to continue my present assignment for a second academic year by which time the pilot stage of the Biology Program should have been completed.

In his endorsement of the request, the head of the science department wrote:

If LTJG McDaniel is not extended, a replacement will be needed, either civilian or military. Hiring a civilian professor takes some time and planning and requesting an officer replacement would likewise present timing problems. Both courses of action are conceivable alternative plans.

Insomuch as LTJG McDaniel is a capable instructor with genuine enthusiasm for his work here, it would seem advantageous to the academy to have him continue for a year to allow firm plans to be made for the long-term future of the biology courses.

Meanwhile, my second year of marriage could not have been better. Mary and I had easily adjusted to our jobs and life in Annapolis. San Diego was a fine place to live, but too spread out and overpopulated, even in the late '60s. Annapolis warmed our hearts and raised our spirits. We were young and newly married, and life was filled with bountiful opportunities.

Early in the morning of September 16, 1969, we had an unexpected knock on the door. The officer next door, who worked in the superintendent's office, said he had seen paperwork indicating that someone else would be moving into our house in two weeks. Amazing. Apparently, I was being released from active duty and the housing unit had reassigned Quarters M even before I had been notified. Congress had cut the Navy's budget, and to save money, the Navy was releasing all reserve officers in the continental U.S. who had less than a year left of active duty. My name was not included on two lists that had been released in August, so I believed I was being retained. The third list carried my name.

We were two weeks into the semester, a poor time to find instructors. The head of the science department had submitted my extension request to the superintendent's office for endorsement. One way I could stay on was to be hired as a civilian instructor. Over the next few days, I had several exchanges with the superintendent's office and was offered a one-year civilian appointment. I countered with a willingness to stay if given a two-year appointment—standard civilian appointments were two years, and I expected the same.

Meanwhile, Mary and I had been devoting our time exploring other options than the offer of a one-year civilian contract, often while walking the fields and woods of the Naval Station. I ruled out a PhD at the University of Maryland because I had not formed in-depth relationships there nor had they offered me financial assistance. Wesleyan was different. The department had initiated a PhD program soon after I had left, and I enjoyed working with and socialising with Berry—we played handball and squash, and he had been a groomsman in our wedding. I called Berry and asked bluntly, "How would you like to have a new PhD student in your lab?"

He asked, "Who?"

"Me!" I responded.

After what seemed like a long pause, he said, "Sure," and then worked out the details, including a stipend and tuition. If the Naval Academy superintendent turned down my counteroffer, I had a backup plan.

On Friday, September 19, I was in my office pushing paper, waiting for the call. That morning, it came: the superintendent was sticking with his original offer of a one-year contract. I could not believe it. I called personnel and instructed them to cut my orders. That evening I asked our neighbour, "Why did the superintendent turn down an optimal solution in exchange for a crisis?" He replied, "The superintendent thought on such short notice you could not possibly get a better deal." Two weeks after that early morning knock, I was taking notes in Wesleyan's graduate physical chemistry course.

I took up in Berry's lab as if I had never left. I had just a few required courses; mostly I would be doing research to provide material for my PhD thesis. Because I had done well and enjoyed my master's research on hormonal control of insect metamorphosis, I continued this line of research. Pretty quickly, an interesting observation provided me with a straightforward question to tackle. A paper written in 1960 had reported that when chilled pupae of *Callosamia promethea*, a silk moth, were given caffeine, theophylline, or theobromine (all methylxanthines), a permanent arrest of development resulted—metamorphosis was blocked. According to the publication, the methylxanthines caused a degeneration of the central nervous system while other tissues remained unaffected. I confirmed the observation that caffeine arrested development in *H. cecropia* and *A. polyphemus* pupae, but degeneration of the central nervous system did not occur. Simply put, my thesis question became: How do methylxanthines permanently arrest metamorphosis in *H. cecropia* pupae? Although I used several methylxanthines to investigate the arrest of *H. cecropia* pupae, I will refer to them as caffeine throughout for simplicity.

Methylxanthines are nervous system stimulants and the most commonly used drugs in the world. These compounds are found in coffee, tea, chocolate, and other foods. Every morning, millions of people drink coffee for its effect on their alertness, primarily mediated by caffeine. Chemically caffeine is 1,3,7-trimethylpurine-2,6-dione or $C_8H_{10}N_4O_2$.

Caffeine is known to interfere with many cellular processes, including blocking cell division in plants, causing breaks in chromosomes of mammals and plants, interfering with DNA repair enzymes in bacteria, causing chromosome loss in fruit flies, and interfering with calcium ion movement in muscle.

As I learned in my first two research projects, answering a simple question is neither easy nor straightforward. Fortunately, I knew from my master's research how to raise and maintain *H. cecropia*, and I had also learned many of the basic physiological, histological, and biochemical techniques required for the caffeine project.

Earlier experiments had established that caffeine injected into pupae before the second moult of metamorphosis blocked the pupae from developing further. Caffeine has methyl groups attached at three positions around a purine double ring while xanthine is an unmethylated purine that is a normal metabolite in DNA and RNA synthesis; caffeine is not. Caffeine was not cleared from the haemolymph (insect blood) of a pupa; rather, it persisted for months. Xanthine was cleared in less than a week. Thus, caffeine prevents the completion of metamorphosis and persists in the animal's haemolymph.

I perfused saline solution through arrested pupae to remove the caffeine. The pupae did not develop, indicating that caffeine blocks development in a way that does not require its continued presence in haemolymph. Since my first project with Berry, researchers had purified ecdysone (moulting hormone from the prothoracic glands) and cyaserone (plant produced compound that had ecdysone-like activity) and graciously made small amounts available to us. Injecting ecdysone or cyaserone into arrested pupae caused all to break dormancy and initiate development. These results established that the caffeine block was somewhere between the brain neurosecretory cells and the action of ecdysone to promote moulting.

Histological studies of brain neurosecretory cells showed that these cells in caffeine-arrested pupae did synthesise neurosecretory material that was assumed to be brain hormone, but corpora allata (the two glands attached to each hemisphere of the moth brain where the axons of brain neurosecretory cells terminate) did not appear to have released brain hormone. Several months after being arrested by caffeine, the neurosecretory cells and their axons became distended with what appeared to be neurosecretory material. These and the above results indicated that caffeine prevented *H. cecropia* pupae from initiating the final moult of metamorphosis by preventing the release of brain hormone from brain neurosecretory cells. I published two publications with Berry that were essentially the first two chapters in my thesis. Two more publications came from the thesis with further work and assistance from the biology department's electron-microscope technician and an undergraduate working with Berry.

*****

In my third year at Wesleyan I came under the influence of a new faculty member, Peter Carlson, an extremely bright geneticist with an enthusiastic personality who was pioneering the field of plant somatic cell genetics. Somatic cell genetics is the study of genetics in somatic (body) cells as opposed to studying inheritance in diploid organisms produced by sperm and eggs that are germ cells, not somatic cells. In animals, the somatic-germ-cell distinction is clear because the cells that will give rise to eggs and sperm are set aside at the time of embryogenesis, whereas in plants the cells that will differentiate into sperm and eggs are specified much later in development.

Somatic cell genetics became possible first in plants because of the earlier capacity to culture plant cells that were first grown in isolated culture at the turn of the last century. By the early 1940s, plant cells could be cultured continuously in a defined medium, from which entire plants could be regenerated.

Animals and plants are typically diploid, meaning they have two sets of chromosomes—one derived from the father, the other from the mother. A mutation on one chromosome may be hidden by the expression of same gene on the other chromosome of the pair. In a haploid individual (possessing only one set of chromosomes), the expression of all genes is observed. Haploid plants were first produced from pollen of *Datura innoxia* in 1964 and from other plant species since.

Biologists have been able to induce, isolate, and study mutations in bacteria because they have a single chromosome with the bacterium's genes. In principle, haploid plant cells can be employed similarly to bacterial cells to induce, isolate, and study mutations. In addition, cells with a selected mutation can be regenerated into plants in which the expression of a mutation can be studied. For example, cells from a haploid plant can be cultured in liquid media, selected for resistance to high salt concentration, and resistant cells regenerated into salt tolerant plants. Seeds from these plants can then be grown in soil with a high level of salt. This would be plant somatic cell genetics put to the practical purpose of creating a crop plant that can be grown in saline soils. This practical aspect of plant somatic cell genetics appealed to me.

After one year at Wesleyan, Carlson moved to Brookhaven National Laboratory on Long Island. I wanted to do a postdoc in his laboratory and explored that possibility for almost a year. In the end, he could not create a place for me. Carlson had earned his PhD at Yale and suggested I talk with Ian Sussex there; Carlson had done his early plant cell culture studies in Sussex's lab. I renewed my relations with Sussex and described my research on insect development and my recent enchantment with Carlson's vision for plant somatic cell genetics. With genuine interest and enthusiasm, Sussex invited me to join his group; however, like Carlson he had no money for a postdoc.

Mary and I again had a major decision before us. What would I do when I finished my PhD in 1973? Take a salaried postdoc doing something I was not excited about, or take an unpaid position?

*****

Soon after we had moved to Middleton in October 1969, Mary had secured a teaching position in the local schools. Our plan, when we first married, was that she would teach for several years, saving her earnings in a rainy-day account, and then we would have our first child. By the summer of 1970, Mary had been teaching for two and a half years, so we planned for a baby in the summer of 1971, which allowed her to complete another school year. Knowing the average couple took about six months to conceive, we stopped using contraceptives that September. Mary was pregnant immediately! I had heard of pregnant women who said they had only had intercourse once. I did not believe them. Now I do—even more so because Mary quickly became pregnant again when we decided to have our second child.

Middletown teachers were teaching-to-rule—meaning they only do what their contract explicitly specifies—because the teachers association had been unable to secure a contract with the school board. A contract was signed in November 1970; however, to Mary's surprise, it included a pregnancy clause: teachers must resign at the end of their fifth month of pregnancy. Mary arranged to meet with Clem Green, the superintendent, to explain that she had planned her pregnancy and that since she was pregnant before the contract was signed, it would be reasonable for her not to be bound by the clause. Mary explained her position, to which Green replied, "Rules are rules." She would have to resign. Mary reminded him that she had two half-day kindergarten classes, one with 32 students and the other with 34, whereas the contract stipulated that a kindergarten class should be limited to 22. Should not the school district hire a teacher for a third class or increase her pay by 50 percent? "No," Green replied. "That is a different situation."

Green's unwillingness to acknowledge the validity of Mary's position solidified our conviction to challenge the pregnancy clause. She appealed twice to the board of education, first with the head of the Middletown Teachers Association and then with a lawyer she met at one of my naval reserve meetings. Both times the board unanimously supported Green. We had not appreciated that the board really had no other choice; they had hired him to oversee the district's normal course of business.

Response by the parents of Mary's kindergartners was affirming and unanticipated. All but two of 64 parents signed a letter to Green indicating they wanted her to continue teaching. Many came to the second board meeting and gave board members grief about the board's views of pregnant women, who were in fact quite capable of doing almost anything. Many letters to the editor were printed in the *Middletown Press*, as were two editorials and several front-page stories. An op-ed titled *Teaching While Pregnant* summed up the community's response well:

It is surprising that the Women's Lib movement hasn't hopped on the case of the nice lady who is pregnant but wants to go on teaching at Macdonough School. Everybody seems to be for her, but rules are rules.

On the other hand, there are all sorts of rules. At *TIME* magazine, for instance, the girls get three week's pregnancy leave, while in older societies, women sometimes worked right through the harvest, with only a very short time spent during the day for the actual delivery. It does seem to be a matter of personal preference.

Although the board chairman, Mrs Betty Adams, made a valiant attempt to explain the matter on the grounds of planning a replacement, obviously a replacement can be scheduled as easily for the eighth month as the fifth. What really is involved here is the old Victorian concept of keeping fat tummies out of the classroom. We certainly agree with that—the story of life is much too educational a concept to be let loose in any classroom.

Mary had gerbils, cecropia moths, a rabbit, praying mantises, and fish in her classroom—some of which had emerged or given birth. Her principal was unequivocal in his support of Mary; however, he told her that under no circumstances was she to give birth in the classroom.

Meanwhile, Mary had asked the Middletown Teachers Association to bring her case to labour arbitration as permitted by the teachers' association's contract. Articles in the *Middletown Press* motivated a dean at Wesleyan to call Mary and suggest she request legal help from Shirley Bysiewicz, the law librarian at the University of Connecticut.

Bysiewicz volunteered to take on Mary's case as an American Civil Liberties Union lawyer—it was the year of the woman. She was fantastic. Bysiewicz contacted each board member, convincing them to let Mary teach while the case was in arbitration.

The school board and teachers' union picked the arbitrator, a 70-year-old economics professor at the University of New Hampshire. The school board's lawyer had been in Bysiewicz's classes when a student. He merely argued that the contract bound the board to fire Mary, whereas Bysiewicz presented the case in the social context of society addressing women's rights.

At the end of the hearing, the arbitrator asked for someone to drive him to the bus station in Hartford. Mary and I volunteered. We were worried because he seemed old to us, was from New Hampshire, and was likely too conservative to rule in Mary's favour. On the drive, after casual conversation, he commented, "This is an important case. I am going to have to think hard about it." I knew Mary would win!

The arbitrator found the maternity clause to be 'arbitrary' and "resulting in unnecessary financial loss and hardship." He went on to rule that it violated Title 7 of the 1964 Civil Rights Act because it resulted in prescribing a condition of employment. Mary had changed the law in Connecticut, and her successful arbitration went on to be cited by the U.S. Supreme Court in 1972 during a case involving other teachers who had been similarly fired for being pregnant. Mary taught through Friday, June 15. Stuart was born the next Monday, and on the following Friday we brought him to Macdonough. The children's eyes popped out at the sight of their skinny teacher, who had had a huge belly just a week earlier.

*****

Although Mary had planned to stop teaching after Stuart was born, the effort put forth to keep her job made her reluctant to tell the school administration that she would not be returning in September. As the final date to resign approached, she decided to stay on. She was still teaching in Middletown when I completed my PhD in 1973. As we weighed my postdoc options, Sussex's lab—with no stipend—won out. This way Mary could keep her job and we could have a second child without re-establishing ourselves elsewhere.

Additionally, Sussex was one of the rare botanists whose perspective on plants was more developmental than physiological. Physiology deals with the functions and processes of living organisms including their parts and organs, whereas development concerns the events and processes that occur in the making of an organism from an egg to the organism's death and back to an egg. At the time, botany had no focus on plant development, whereas zoology had a major division on development. Needless to say, my background in animal development complemented Sussex's perspective.

3.8: Ian Sussex standing in one of Yale's greenhouses.

Once I had decided on Yale and completed my PhD, I began seeking postdoc funding. I applied to the Research Corporation and National Institutes of Health (NIH) for postdoctoral fellowships and with Sussex to the NSF for a grant titled *Auxin Mutant Selection by Somatic Cell Genetics,* all without success. For my second year, however, Sussex provided me with a postdoctoral position funded by Yale's NIH trainee grant. Mary was pregnant and again taught school through

Friday, June 7. Virginia was born the following Monday. The next day, we brought her to the lab to meet the Sussex crew.

Sussex's group, when I joined, was composed of two faculty members on sabbatical, a research associate, and four graduate students; the following year three more graduate students arrived. During morning and afternoon tea—Sussex was a New Zealander—everyone took part in open discussions ranging from consideration of a research problem someone was having to the chemical composition of coffee. Not infrequently, Sussex walked in declaring, "I have an idea." Our discussions were wide-ranging, unscripted, and informative. At one morning tea, Ed Yeung, a graduate student from Hong Kong, and I decided to fix lunch for everyone. It went so well that thereafter we had spontaneous lab lunches frequently.

Virginia Walbot, who left the lab two years before I arrived, summarised her Yale experience in a booklet for the 1985 reunion of Sussex's students.

We met three times a day for coffee at 10, noon, and about 3. [The] purpose of the coffee club was to dissect people's personalities and also to discuss the structure, reproduction, and evolution of plant form. By my calculations we spent approximately 1,000 hours as a group during the four years I was in the lab. Considering that most courses are only 40 lecture hours, the coffee hour was equivalent to 25 botany courses.

Jim Wong, who worked in the lab with me, wrote:

I started working in Ian's lab eleven years ago. It is appropriate to say 'Ian's lab' rather than 'with Ian' because… a remarkable group of trainees assembled under Ian's benign guidance, and it was my interaction with this group which was the basis for my graduate education… We were at least partially aware of the unique talents in that group, and even from this short perspective, the accomplishments and continuing promise of its fellows are impressive. How could plant developmental biology support all of us, I wondered? Well, it is clear now that we have become a major factor in this field, permeating and probing most of its current research areas and commercial applications. It was my tremendous good fortune to fall in with such colleagues and friends.

Ian's personality and pedagogy of course set the tone that made all of that possible. A liberal education, diversity of interests, and approaches—these are the objectives that somehow, I absorbed. Tools of the trade? Perhaps, but many students graduate without them. Thank you, Ian.

The reality of Wong's assessment of Sussex's students was exemplified in 1988 at the national UCLA Plant Development Meeting in Steamboat Springs, Colorado. Eleven of Sussex's PhDs and postdocs attended, with six giving invited talks. Some colleagues kindly referred to the group as The Sussex Mafia. Yes, beginning in the 1970s and into the 1990s, plant developmental biology replaced morphogenesis as a major area of botany, with many from Sussex's lab playing important roles. Thirty PhDs and 27 postdoctoral fellows enhanced their understanding of science and life under the influence of him and those in his lab.

My Sussex group education began in the summer of 1973. I learned and applied the techniques of plant tissue culture and somatic cell genetics to induce, select, and analyse plant hormone and amino acid analogue resistant mutants in haploid fern spores and haploid *Nicotiana tabacum* cells. You might wonder, "Why tobacco?" Tobacco at the time was a model system, because the plant was easy to grow, haploid plants could be produced from cultured anthers (the male part of a flower), cells and tissues could be grown in culture from which plants could be regenerated.

N. tabacum cultivar Wisconsin 38 was the tobacco employed in the Sussex lab. This cultivar takes three to four months to progress from seed to flower. The time period to get anthers at the right stage for producing haploid plants can be substantially shortened by cutting off the top of a plant that had flowered, thereby causing the growth of one or two axillary buds—buds located at the base of a leaf where the leaf joins the stem—that then flower relatively quickly compared to a seed derived plant.

When employing this shortcut, I observed, as many had before me, that axillary buds at the top of the plant produced few nodes—the attachment site of a leaf to the stem—before flowering, while buds lower down on the stem produced a greater number. It appeared to me that an axillary bud meristem (the cells at the tip of a shoot that divide to produce the stem and its structures) flowered after producing the same number of nodes that had been above it on the portion of the stem cut off. Similar to the mature, above-ground, tobacco plant—leaves and stem as well as the terminal flower—is produced by the shoot apical meristem. This made me think of animal development in the case of marine annelid worms that grow back the same number of body segments cut off; somehow the regenerating tissues counted segments. I wondered: How did a tobacco meristem count the nodes it produced? It would take almost a decade before I understood that I had stumbled onto the question from which my major contributions to science would come.

To put this node-counting question in a larger context, one needs to know that the development of a multicellular organism is a sequential process, as I mentioned earlier when discussing my lack of depth perception. The problem I was trying to resolve is a central developmental question for any structure that has a growth period that ends. Our arms began as limb buds that cease growth with the formation of a hand. Likewise, the above ground tobacco plant begins as a vegetative shoot apical meristem that produces stem, leaves, nodes, and then ceases growth with the formation of the terminal flower. That is, when the flower is made, the number of nodes is fixed. How is the meristem instructed to cease node production and make a flower? That is the fundamental developmental question I was asking.

In the formation of these final structures—hands and flowers—a sequence of commitments to various developmental fates are made that may or may not be reversible. That is, how flexible is the developmental program? Comparatively speaking, commitments are fixed in animals and unstable in plants, but not always absolutely fixed or totally unstable. The dice rolled favourably for me when Sussex's lab used a particular horticultural variety or cultivar of tobacco that not only counted nodes precisely, but also ended node production with a relatively stable determination state that preceded flower formation.

Chapter 4
Doing Science

We recently called our granddaughter on her birthday. After we sang Happy Birthday to her, she announced definitively, "I'm 5!" How did she know this? How did she count the years? I was in awe. It was an amazing accomplishment to which we give little thought.

A few years ago, brood V of the 17-year cicadas that are composed of three species (*Magicicada septendecim, M. cassini,* and *M. septendecula*) emerged in Oberlin, Ohio, and its surrounding communities. This emergence of periodical cicadas was the first I had experienced. For several weeks, a constant, loud noise emitted from seemingly all directions, similar to the incessant roar of a major highway. The cicadas' tymbal organs, a pair of ribbed membranes at the base of the abdomen, produce a noise that reaches 120 decibels, about the intensity of a jackhammer close-up.

While sitting on our deck, I could see dozens of cicadas flying this way and that. They are big: a quarter inch in diameter and over an inch long with two big red eyes that bulge from their heads. Cicadas alighted on most of the deciduous plantings in our yard: buckeye, maple, oak, and river birch trees, as well as lilac and snowball bushes. Every time I passed the lilacs behind our barn, I would catch a few and toss them into the pond. The instant a cicada hit the surface of the pond; it was gone—the fish loved them.

After mating, a female uses her ovipositor at the rear of her abdomen to slice into the bark of twigs at the end of a branch and deposits eggs. About 400 eggs are laid in dozens of twigs. Many of these twigs die, setting back small trees; larger trees are not fazed. The eggs hatch in several weeks. The nymphs drop to the ground where they dig into the soil. For the next 17 years, they live on sap from tree roots before emerging once again. Scientists are still puzzling out the biological clock cicadas employ, not an easy task when nymphs take 17 years to mature. Some evidence indicates that nymphs may sense something associated with each time a tree produces a new batch of leaves as a means for counting years. Size, too, may be relevant because most species of cicadas are not periodic and emerge from the ground several years after nymphs complete their fourth and final larval moult. However, no one knows how nymphs decide to burrow to the surface after 17 years, when they then climb a tree and moult into adults. Nevertheless, we do know the 17-year cicadas are among the longest-living insects, with termite queens holding the record of at least 50 years, perhaps as many as 100.

Tobacco plants, too, are very good at counting, not years but nodes. *N. tabacum* cultivar Wisconsin 38 produces a uniform number of nodes. Other cultivars of *N. tabacum* as well as other *Nicotiana* species flower after producing a uniform number of nodes. All of these *Nicotiana* species cease producing nodes when the terminal meristem stops making nodes and develops into the terminal flower.

How does a tobacco plant count? I talked about this question with my Yale officemate, a Sussex graduate student. After several weeks of talking, he encouraged me to try something. We discussed various experiments, which I refined and then conducted. First, I confirmed that a population of 30 tobacco plants produced a uniform number of nodes: 36 with a range of 34 to 37 nodes. I then showed, at the time the plant flowered, that axillary buds at the base of the plant produced more nodes than those in the middle of the plant, which in turn produced more nodes than those at the top of the plant: bud at the base, 39; bud in middle, 23; bud at top, 14.

These results clearly established two things. First, the number of nodes produced by a tobacco meristem is quite uniform. This was not new. Second, the number of nodes produced by the shoot apical meristem of an axillary bud, when forced to grow by removing the stem above it, is position-dependent. This, too, had been observed earlier; however, the focus on being position-dependent opened new avenues of thought about plant development not readily apparent from just a physiological perspective. These observations permitted me to imagine that tobacco shoot apical meristems count nodes as the mechanism by which a shoot apical meristem is instructed to make its terminal flower. This developmental perspective was unusual in botany and plant physiology.

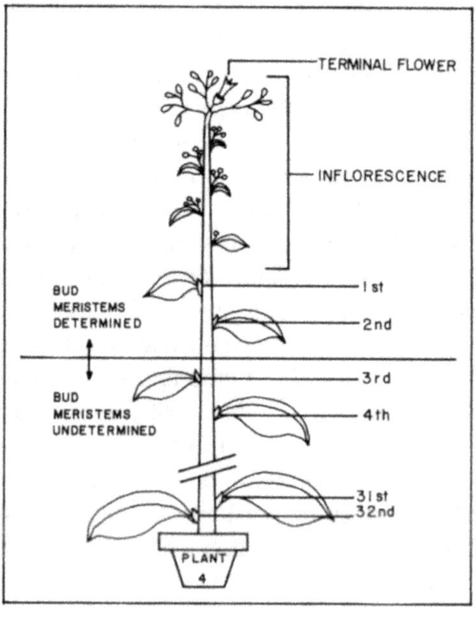

4.1: Labelled Nicotiana tabacum plant.

Let me describe the architecture of a mature tobacco plant so you can appreciate better the decisions I made to explore node counting. An axillary bud meristem produces seven to nine leaf primordia and is then arrested via a process called apical dominance. This results in a plant with a single stem with leaves and arrested axillary buds. Above these buds are three to five floral branches that grow out, producing flower after flower. The terminal meristem, which has produced leaves, nodes, internodes (the stem between nodes), axillary buds, and floral branches, ends its existence by differentiating into the terminal flower. My studies were of developmentally arrested axillary buds and of terminal buds before they formed a flower.

I employed rooting and grafting of buds to further establish that it was not the initial position on the stem that correlated with the number of nodes produced by a bud moved to a new environment. Twenty-five buds from various nodes were rooted and grown to maturity. They produced 36 nodes, indicating that rooted axillary buds had no memory of their initial position on the main axis. When buds from the upper part of the stem were grafted to a similar position on another plant and released from dormancy by decapitation of the plant above the grafted bud, they produced 25 nodes, while when grafted to the base of another plant and released from dormancy, they produced 37 nodes. When basal buds were grafted to the base of another plant and released from dormancy, they produced 41 nodes, while when grafted to the upper part of another plant they produced 27 nodes. Again, it was the grafted position of the bud on the stem that correlated with the number of nodes produced; more when in a basal position and fewer when in an apical position.

These experimental data supported the hypothesis that tobacco meristems have a position-dependent means of precisely counting nodes, or something tightly correlated with nodes. Sussex, too, was keen on this position-dependent phenomenon and in September 1975 arranged for me to present a paper at a symposium on juvenility in plants at which he was speaking. I was also writing the results for publication in a journal. My side project in Sussex's lab had yielded publishable results, although my major interest in somatic cell genetics had yet to yield similarly important and publishable data.

All the while, I continued seeking a permanent job. My teaching experience at Annapolis had solidified my desire to be a college teacher. Unfortunately, teaching jobs were scarcer than postdocs. I was hesitant to consider a research-teaching institution where research was all that counted—where faculty were expected to publish often in quality journals and secure grants to finance research and support graduate students. I envisioned finding a position in a teaching-research school where I could work in collaboration with high-profile scientists like Carlson and Sussex yet have my tenure decision heavily weighted on teaching and service, not research. Personal desires are fine, but the stark reality of finding a quality job in higher education is simply stated—apply to every job for which you are qualified.

Since arriving at Yale, I regularly scanned the jobs-available section of *Science* and followed up on the rare one for which I was qualified. In November

1973, I saw an ad in *Science* from Tom Scott at the University of North Carolina; he was the Oberlin faculty member for whom I was a senior research student seven years earlier. I wrote him a long letter about what I had done in those years and expressed my interest in the position he was filling. Scott wrote, "I must say I find real gratification and satisfaction in hearing such great news about your postdoc with Sussex." Scott was happy to ask the search committee to consider my application; the developmental physiologist position fit my qualifications to a tee.

I submitted a formal application in November. Scott wrote in February that the committee had completed its preliminary but thorough screening of "a very large number of applicants" and that "we do not foresee that we will be able to act favourably on your candidacy. I wish to assure you that this decision was reached after very careful deliberation." I was shocked and disheartened—I had not even survived the first cut.

Earlier, when I was seeking a postdoc, Bill Sheridan at the University of Missouri-Columbia had set up a postdoc for me. I familiarised myself with Sheridan's department. In spring 1974, Peter Carlson informed me that Sheridan had not received tenure and that a person doing plant tissue culture and somatic cell genetics would fill his position. I wrote to the director of the biological sciences division to express my interest. He replied, "It is obvious that a person with your background would be an asset to our division," and that they would be selecting interviewees in the fall. I had a two-day interview in early December, after which the director wrote: "I want to express my personal appreciation for the stimulating seminar you presented recently. It was the kind of high-quality discussion which we like to have for our faculty and students."

A few months earlier, in October 1974, I had responded to a *Science* ad for a developmental biologist position at Rensselaer Polytechnic Institute in Troy, New York. Walter Auclair, head of the search committee, called at the end of October to ask if I could come for an interview on November 20. I had a conflict and pushed the interview to December 11. This was fortuitous. I went from being the first candidate interviewed to the last. Anecdotal data indicate that interview sequence affects the outcome, with the last candidate being selected more often than the first. Auclair called me on December 18 to alert me that my reference letters were incomplete—Sussex's letter was missing. He sent his letter the next day. A few days later, Auclair offered me the position.

Mary and I drove to Troy to meet people and consider the job. When talking with Auclair, he told me that they had had two excellent candidates and had a hard time choosing. He continued, "Sussex's letter clinched the offer for you." Sussex had given me a copy of the letter in which he set a very high bar for me to achieve.

Carl is an excellent researcher. He has the ability to ask clear questions and to design experiments that give clean answers… He has done research on auxin physiology of peas, enzyme biochemistry of fungi, and regulation of diapause in insects, and he therefore brings to his research a breadth of experiences that is rare among people of his age. In addition, he is one of those people who are not satisfied with a superficial answer, and he works at a question until he understands the basic premises that underlie it.

I believe that somatic cell genetics and related developmental studies will become a major emphasis of plant research in the next few years, and I think that Carl has the training, the background, and the ability to be one of the leading figures in this field.

Rensselaer in 1975 emphasised teaching while expecting some research. A biology faculty member, an accomplished teacher who had published just once, was told he would have been granted tenure if he had had one more publication. I knew I could clear that research bar. At the same time, I had been seduced by my research experience at Yale, which made the University of Missouri position attractive; I had had an offer there that hinged on a hiring freeze being lifted. The hiring freeze held. I accepted the Rensselaer offer with enthusiasm!

Going to Rensselaer meant my time at Yale was drawing to a close. I loved being a postdoc in the Sussex group for a host of reasons. It was a time of true freedom to pursue whatever came up and engage with exceptional scientists in the making. The atmosphere was conducive to growth and creativity primarily because Sussex made any question or idea a good one. He warmly accepted me and considered my knowledge of animal development an asset. I felt no constraints; I could do anything I wished. Sussex gave everyone lots of running room. If advice was not sought, Sussex rarely gave it. Alternatively, when advice or time to talk was needed, he fit it into his schedule. Sussex set the tone for professional and personal affairs that others and I sought to emulate. I woke up each morning wanting to do science.

Everyone in the Sussex lab had an independent project; this was atypical for university biological research laboratories at the time. We were exposed to a diversity of topics in the mid-1970s that played a major role in the education of Sussex's students—both postdoctoral fellows and graduate students—who in the coming decades would make significant contributions to many areas of botanical science and science itself.

I thanked Sussex for welcoming me into his lab and for his unfaltering support, acknowledging what his group had done for my botanical education and for me personally. To my surprise he expressed his appreciation by stating that my presence in the lab had been a stabilising force over the past two years because serving as dean of graduate students had limited his time in the lab. A few days later, when I left my postdoc for good, I cried while driving to Middletown. My professional identity would be forever linked to the Sussex lab.

My pleasant postdoc life at Yale was replaced by an overwhelming challenge: setting up a lab while simultaneously developing a viable research program, securing research funding, teaching classes, and help settling our family in Troy. In February, I signed a three-year letter of agreement with Rensselaer for a nine-month salary of $13,000 plus moving expenses of $500. The department provided me with $6,000 for equipment and $3,500 for expenses to set up and initiate my research program.

My teaching assignment for the fall semester included a graduate course on development co-taught with Auclair, one-third of the senior-projects laboratory course, and lectures on photosynthesis in introductory biology. Although I was hired as the department's developmental biologist, the course planners thought a plant researcher would naturally know enough about photosynthesis to teach it, but the last time I had thought about photosynthesis was in 1961 during introductory biology at Oberlin.

Similar to what I was told at Oberlin's first-year orientation—look to your left, look to your right, in four years one of you will not be here—Joe Landau, the department chair, told me once I was settled, "We'll put many things in your way that you will have to overcome to be successful." I had been forewarned.

A new, wet-behind-the-ears, university faculty member is faced with an intractable set of simultaneous, long-term projects. Consider an accomplished juggler who eventually achieves a steady state with half a dozen or more objects in the air. She does not toss them all at once; rather she launches them one at a time while keeping the others within her grasp to get them airborne as quickly as possible.

Before leaving Yale, I had submitted a manuscript for a paper to be presented at a Symposium on Juvenility in Woody Perennials and subsequently published in *Acta Horticulturae.* I also submitted to *Nature* the same data streamlined to meet that journal's specifications. In late September, *Acta Horticulturae* accepted the paper. However, *Nature's* editor wrote, "... in our opinion your paper is of insufficient general interest to complete successfully for our limited publication space." Enclosed were comments from a reviewer.

I called the editor, who listened to me explain why I thought the paper important and of general interest. He suggested I revise the first paragraph to clearly state why the topic was important and what the data established, and then conclude the manuscript with a similarly focused statement. I was given a second chance. The two reviewers of the revised paper had no substantive comments, and the paper was accepted. This was the closing paragraph:

This regulation of pattern development may be analogous to the model of animal development, where differential gene expression is thought to be intimately related to the orientation of a cell within a developing structure or organism, and that this orientation involves the reception and interpretation of a positional information signal. Although the model has not often been used in studies of pattern development in plants, positionally-related development of shoots and leaves has been observed in numerous plants. These observations and the above results suggest that the use of a positional information model could introduce new ways of examining and interpreting developmental patterns in plant systems.

*****

When I left Sussex's lab, I had had a number of experiments underway in Yale's greenhouse, and Sussex kindly allowed me to continue them, offering staff support for watering and starting new plantings. My friend Ed Yeung, still a graduate student, kept an eye on the experiments and helped as needed. When the time came, I served as a job reference for him.

The experiments at Yale extended our earlier observations. In my published papers on node counting, I examined the developmental potential of axillary buds located four nodes below the inflorescence, which is composed of four or five floral branches. To assess growth potential of the uppermost axillary buds, I rooted the top five from several plants and allowed them to mature. Numbered from the top down, these buds from one plant did the following: bud 1 made 18 nodes; bud 2 made 17 nodes; bud 3 made 37 nodes; bud 4 made 38 nodes; and bud 5 made 38 nodes. When examining the growth pattern of the upper-most axillary buds from many plants, we noticed the separation between few-node plants and many-node plants exhibited a step-function that always occurred between buds 1 and 2, buds 2 and 3, or buds 3 and 4 (Illustration 5.2). I then grafted the upper five axillary buds from individual plants to the base of other plants. The same one-node separation between few-node and many-node shoots was observed.

4.2: Nodes produced by rooted buds taken from the same plant. The line separates determined buds from undetermined buds for that plant.

Number of Nodes Produced by Rooted Buds from the Same Plant

Bud Number	Plant 1	Plant 2	Plant 3	Plant 4
First	15	18	19	18
Second	18	17	38	17
Third	20	37	38	37
Fourth	38	38	38	37
Fifth	40	38	39	38

These results established that the uppermost axillary buds that produced few-node plants when rooted or grafted were developmentally determined for this few-node growth pattern. That is, if a cell, tissue, or structure exhibits the same developmental fate in place, when moved to another place on the organism, and when grown in isolation, it is developmentally determined for that fate. This concept of determination had been extensively used and studied in animal systems, but not in plants. In October 1977, I submitted a manuscript to *Developmental Biology,* stating in my submission letter:

[This paper] is concerned with determination in a plant system and is one of the few plant examples of determination using the [animal developmental biologist's] criteria of growth in isolation and growth in a different location on the organism [to demonstrate determination]... Plant researchers in general do not think in terms of determination in plant development.

I made this statement because the developmentally based grant proposals I had submitted to NSF in 1975 and 1976 had been rejected; most of the reviewers had made comments revealing their strong physiological perspectives and a limited understanding of my development approach. For example, these four physiological-based statements were given in assessments of these proposals: 1) "... instead of investigating the physiological-biochemical basis of this phenomenon." 2) "The existing findings are interesting, but might perhaps be differently interpreted." 3) "No consideration is given, for example, to possible effect of nutritional competition, which is known to have a considerable effect on bud outgrowth in flax." 4) "... in terms of apical dominance."

This problem made me recall the difficulty people had when helping me to estimate the number of cecropia larvae growing on a netted cherry tree. If they had never seen a larva, they did not know what to look for and were unable to see any although many were present.

In December, the editor replied, "I regret having to write you that we shall not publish your manuscript in its present form." Sure enough, the reviewers were plant scientists who did not understand determination or appreciate its importance in plant development. The primary reason for this, I believed, was that some plant cells taken from a mature structure have the capacity to regenerate an entire plant. Thus, how could the cells be determined for a particular fate, if they can express all fates?

The reviewers did correctly criticise some misstatements and bogus ideas presented in the manuscript. I addressed them and submitted a revised manuscript; it was accepted in May and published in September 1978.

I sent a copy of the paper to Anton Lang, director of the Plant Research Laboratory at Michigan State University. Lang was a respected, well-known plant scientist who helped formulate the hypothesis that flowering was controlled by a hormone. In his reply, he said he hoped I found Rensselaer's facilities and environment satisfying, adding, "Quite frankly, I have not considered this institution as a centre of **plant** biology, but trust it will become one thanks to your efforts." This was quite a compliment. Lang was a member of the National Academy of Sciences and editor of *Planta*, a prestigious plant journal.

When I interviewed there in 1974, Rensselaer was a teaching-research university with a low research bar for publications and no hard requirement for obtaining external funding. Unbeknownst to me, the board of trustees in the mid-1970s resolved to significantly enhance Rensselaer's research profile. The new president, George M. Low, had run the NASA Apollo Mission that put men on the moon and was hired in 1976 to transition Rensselaer into a major research university. Although I recall no explicit announcement, I sensed during my first year that research would become the criterion on which my tenure would hinge. Perhaps that was what Landau implied when he said, "We will put many things in your way that you will have to overcome to be successful."

During one fall evening in 1975, while getting a bath from Mary, Virginia spoke her first sentence: "Daddy at work." Adrenaline fuelled my first year at Rensselaer, replaced in my second year by unrelenting stress. My third year seemed similar to the middle miles of a marathon—just keep pace. I worked every day, including weekends, often returning to the lab after dinner. Some nights Mary and the children brought dinner to the lab so I could work into the night with few distractions. On weekends, I would bring the children to the lab, where they enjoyed pipetting water from one beaker to another or drawing at my desk.

During my first three years, I taught two courses each semester, wrote several request letters and two proposals for a department greenhouse, organised and ran a program to acquaint graduate and undergraduate students with faculty research, and served on the department's graduate qualifying exam committee. I submitted four research grants—two to NSF and one each to the National Aeronautical and Space Administration and the Herman Frasch Foundation—without success. However, my lab produced data for two research papers that were in press the spring of my third year.

In early spring that year, I felt the rapidly rising bar for tenure when the provost denied me a second three-year contract. I was surprised! Given my accomplishments, I had assumed the contract was a sure thing; neither Landau nor anyone else had warned me otherwise. Landau went to bat for me. The provost eventually consented, but with a caveat: "Tell McDaniel if his research is not externally funded, not to even think of applying for tenure." There was no doubt about it: Rensselaer's research expectations had increased dramatically. My job at a teaching-research university had morphed into a research-teaching job that I was not sure I could manage.

Prior to this controversy, my lab, with contributions from two senior theses and data from Merrill Blackman, my first graduate student, had produced sufficient preliminary results to submit three grants for our somatic cell genetics projects. In July, NSF awarded us a small grant for shaker equipment; NIH provided a three-year grant of $143,855 for amino acid transport systems and mutants; and the U.S. Department of Agriculture awarded us a two-year grant of $20,000 for the isolation of herbicide resistant mutants.

Meanwhile, Mary and I had planned the family's first vacation in six years. We would meander and car camp across the country to Calgary, Canada, where I would present a paper on amino acid transport in cultured cells at the Fourth International Congress of Plant Tissue and Cell Culture. For a month, we explored the Badlands; Mt. Rushmore; Lusk, where we stayed with the Bardos; Wind River Mountains, where Virginia rode her first horse, and which almost threw her when spooked by the bang of a rifle shot; and then onto Grand Teton, Yellowstone, and Glacier national parks. We also enjoyed a stimulating reunion with Ed Yeung, my Sussex colleague who was now teaching at the University of Calgary. While Yeung and I attended the conference, his wife showed Mary and the children the sights and introduced them to a diversity of Chinese cuisine. Our trip home was fast—a 32-hour drive to Wauwatosa to visit Grandmother McDaniel and then home to Troy.

Now, with new funding from the three agencies, my lab went from rags to riches overnight and would stay funded for 20 years. The NIH grant provided us $10,000 to renovate Rensselaer's greenhouse. However, upon thorough examination, our facilities personnel determined that the shell of the greenhouse was in poor condition and needed replacement. In June 1979, per standard university procedures, we submitted eight requests for proposals. We received three bids: $19,800, $22,000, and $29,000. The lowest bid was acceptable and included maintenance and improvement costs: $8,360 and $11,440 respectively. Rensselaer covered maintenance costs—$3,000 from the School of Science and $5,360 from Physical Facilities—and the NIH provided improvement expenses. The enhancements to the greenhouse provided quality-growing space, meaning I no longer needed to drive to Yale once or twice a month.

At Yale, I had initiated multiple lines of research because it was impossible to discern which would be fruitful. I continued the most promising three projects at Rensselaer. Good fortune shone on us when all of them yielded publishable results and recognition by the early 1980s. In 1979, I was invited to the international Plant and Tissue Culture Gordon Conference to speak on positional information in axillary bud development. In 1982 I spoke at the New York Botanical Gardens Symposium on Plant Tissue Culture: New Trends for the Horticulturist, and in 1983 I was invited to write a review article titled, *Transport of ions and organic molecules* for the *Manual of Plant Tissue Culture Vol. I. Techniques and Applications*.

Recognising my limits and those of our small operation, I faced a conundrum; it was clear to me that we could not continue to successfully pursue all three. Yet, by initially doing all of them enabled me to accomplish three important goals for a scientist (in order of importance): recognition by plant physiologists and botanists; a substantial increase of peer-reviewed publications in well-respected journals (eight to 18 from 1975 to 1982); and to attain tenure. Edward O. Wilson, among the most eminent biologists in the second half of the 20th century, rationalised this order of importance.

... to be a real scientist... is to make a certifiable scientific discovery. The acid test among professionals is to be able to finish this sentence: "I discovered..." The importance of the discovery is judged by peers who dwell in the same silo and nearby silos. (Edward O. Wilson, 2017, *The Origin of Creativity, pg. 191*)

Many researchers could do herbicide selection and amino acid transport. I, however, was among a few who could bring an animal-developmental-biology perspective to plant science; this was also what interested me the most. The choice was clear and affirmed when in 1982 I was awarded a two-year NSF grant on the "Stability of the Determined State." The story behind this grant reveals the role of serendipity in all our lives.

In January 1980, Mary and I travelled to San Diego for my brother's wedding. While there, I arranged with Susan Bryant, a developmental biologist who studied pattern formation and limb regeneration in the axolotl (Mexican salamander), to give a seminar titled, *Regulation of Determination in the Tobacco Shoot Meristem* to the Department of Developmental and Cell Biology at the University of California-Irvine. Most of the faculty there researched animal systems; a few studied plants. I had met Bryant at meetings of the Society for Developmental Biology and was excited to present our findings to people who thought developmentally.

I spoke for 45 minutes on the contents of three publications—position dependent development (*Nature*, 1976), determination for growth pattern (*Developmental Biology*, 1978), and the influence of roots and leaves on meristem development (*Planta*, 1980)—as well as the questions arising from these data and conclusions. For the next 45 minutes, we had an animated discussion that Bryant interrupted to allow people to leave. A dozen left. Over wine and cheese, we continued the discussion for another half hour or so. I was on cloud nine! It was the best seminar I ever gave, before or since. More than 60 developmental biologists were engaged in high-level discourse on a central question of the discipline: How is a group of cells in a limb bud or a meristem fated to become a hand or a flower? Mary was in the audience, witnessing science and her husband at their best.

A year later, Bryant became the director of developmental biology at the NSF and reviewed my new grant submission on the phenomenon of determination. Three outside reviewers ranked it 'very good'; the fourth ranked it 'good'. This was a huge improvement over my two earlier NSF submissions, which earned 'fair' rankings and one 'excellent'—in my biased assessment that was the only reviewer to actually understand the proposal. With this latest proposal, I had still not convinced the reviewers of the importance of characterising in a plant system the commitment step to a specific fate, in this case 'make a flower'. The summary read:

The panel recognises an intrinsic value in the experiments proposed—they can lead to some very interesting information about floral induction and research in that area has been bogged down for a long, long time. The initial experiments are carefully designed. However, the panel feels strongly that the molecular work should be omitted at this stage of the research... The panel is impressed with the initial experiments and recommends modest funding for one year to carry these out.

At this time, the NSF was funding about one of every six proposals; my proposal fell in the grey zone just below the cut-off point. Had Bryant not hosted the seminar I presented at Irvine, my proposal likely would have been rejected. At Irvine, I had had the opportunity to demonstrate the quality of our work and establish why a developmental perspective was sorely needed in plant research. In the end, Bryant told me I had not made the cut-off, but that she had put together $80,000 for two years support. She closed by saying, "This is not what you need, but something. Now show us what you can do."

This grant came at a time when our work on amino acid transport and herbicide resistance was being supported and yielding publishable research that undergraduates were able to do. Susan Rundell Singer, who had come to Rensselaer to be an engineer, changed her major to biology and became my advisee. Today, she is the provost at Rollins College in Winter Park, Florida. Her illustrious career began in our laboratory in 1977, where we first became colleagues. We remain so to this day. Singer's father was a Rensselaer-educated engineer who had worked for General Electric until he died in an airplane crash while returning from a meeting with Admiral Rickover in Washington, D.C. A teenager at the time of her father's death, Singer had three younger siblings and came to assume significant responsibilities in her family.

Quiet, self-effacing, and extremely competent, Singer quickly expressed interest in what our lab was doing. I had secured samples of amitrole, diphemamid, glyphosate, and picloram as possible herbicides for which to select resistant mutants. We chose amitrole and glyphosate because they appeared to degrade in soil and be nontoxic to animals. Within a year Singer was selecting and characterising amitrole resistance in cultured tobacco cells and regenerated plants. Her undergraduate thesis—Isolation of Herbicide Tolerant Plants—won the 1902 Prize for Best Institute Undergraduate Thesis. Singer's accomplishments, in large part, were responsible for a three-year extension of our USDA grant. We requested $231,943 and were granted $20,000 per year for three years—an annual doubling of our first USDA grant. How wonderful!

I was granted tenure in January 1981 and became eligible for a semester-long sabbatical at full pay. With NIH funding a full-time technician and Singer in the midst of her master's thesis, I was able to take the sabbatical, knowing our research would continue unabated. My preferred destination was the University of California-Davis because the excellent plant scientists there would enhance my botanical education. I had been in conversation with Wesley Hackett in the Department of Environmental Horticulture about his flowering studies of *Bougainvillae* 'San Diego Red' and *N. tabacum* Wisconsin 38. I would learn more broadly about flowering by spending time with his group and with his colleague, Roy Sachs, who had been introduced to flowering by Anton Lang.

Finally, I needed a break from the intensity of my seven years in Troy, as did our family. After all, 'sabbatical leave' is defined as "a year or half year of absence for study, rest, and travel, given at intervals, originally seven years." We planned to take a leisurely month driving across the country and a similarly relaxed trip home.

*****

In the midst of sabbatical planning, Peter Carlson from graduate school days called. We were in frequent contact, and our families had spent time together. He had divorced and moved from Brookhaven National Laboratory to be the John Hannah Distinguished Professor of Agriculture at Michigan State University, where I had travelled to give seminars, to help with his second wedding, and to lend a hand on his farm. He knew I had just earned tenure and was planning a sabbatical to Davis, yet he had a surprising request. Would I join him as chief scientist for the company he was forming?

The company would employ tissue culture techniques to mass-produce disease-free sugarcane and banana plants. Carlson had teamed up with John Henry, a Harvard graduate with a law degree from Columbia who was working for a law firm in New York. Carlson sent me a draft proposal, and we arranged to meet with Henry in New York City in April. Carlson was an exceptional scientist eager to use science for the betterment of humanity. We got along well, and I was motivated by the opportunity to work with him again.

We met at Henry's law firm to write the business plan. Sugarcane and banana were ideal crops because they are propagated from cuttings, with the potential 'seed' market valued at billions of dollars. Cutting a cane into pieces with one or more internodes and planting the cuttings in soil is the conventional procedure for propagating sugarcane. This allows the roots to form from the base of a node where the leaf attaches and for shoots to grow from axillary buds. Viruses build up in apical meristems over time and significantly reduce plant vigour. Regenerating shoot meristems via sterile culture of explants from superior canes produces unlimited virus-free plantlets that grow into canes, which in turn are used to make internode-node pieces to plant. Similar procedures are possible with bananas.

For the next three months, we met almost weekly in Henry's conference room. The meetings lasted an entire day and usually ran into the night, with us writing, researching, and rewriting until we could do no more. Henry had a hard-driving personality; Carlson and I called him 'John Henry, the steel driving man' after an African American folk hero who, in a contest with a steam-powered steel driving machine, beat the machine by driving the steel rod 14 feet into rock to the machine's nine feet.

In June, Carlson's brother, Bill, a glass artist, had a show in New York City. We all attended, Mary included, and proposed that Bill create an art piece for launching Plant Genetics. It was becoming a family endeavour.

On Sunday, June 28, *The New York Times* ran an article in the business section, "The Gene Machine Hits the Farm," written by Ann Crittenden, Henry's girlfriend. She commented on the increasing number of major companies that were joining the plant biotechnology bandwagon—Frito-Lay, Campbell's Soup, Du Pont, Pfizer, Monsanto—as well as a few start-up biotech companies such as Calgene and Agrigenetics. The story's sidebar, titled *A Campus Brain Drain,* led with this:

This spring, the 39-year-old plant biologist finally won tenure at his university in upstate New York. After 15 years of teaching and a respected research record, his salary was still $24,000, a sum that the university more than recouped through its share of his outside research grants. When the chance came last month to join a friend in a new start-up company, for almost double the money, he jumped at it. He is breaking the news to the university this month.

I was the biologist, of course, but the piece had a few misstatements. The business plan for Plant Genetics, Inc., was in good draft form at the end of June: Henry as president, Carlson as vice president for research, Lou Nickell as the first member of the science advisory board, and me as senior scientist. Nickell, a senior citizen, was an accomplished scientist. He served as an officer in the Marine Corps during World War II and then as vice president for research and development at Velsicol Chemical Corporation. He had also chaired several international plant tissue and cell culture conferences. We gathered in New York on July 1. Nickell and I met to discuss a list of topics, from virus selection and downy mildew resistance to his work in Hawaii selecting for increased yields in sugar cane.

Henry and I arranged to meet the following Monday. I took the early train from Albany, but Henry never showed up. This was the second time he had stood me up. On the ride home, I began contemplating the situation between Henry and me. Bad vibes had been in the background from almost the beginning. Mary had sensed them at Bill Carlson's art show but did not say anything at the time.

Henry's value system and approach to people conflicted with mine. I respected people and strived to create a team that cooperated and built trust, with a free flow of information. He had made no effort to develop a working relationship with me based on respect and trust. He was a steel-driving man, and I was the drill he was using to accomplish his objectives. Henry needed Carlson; he did not need me. Carlson would need to act as a buffer between Henry and me if we were to go forward. But this was a bad position in which to put Carlson. I knew it would not work in the long run.

The next day, I called Carlson for a heart-to-heart talk. We both agreed that my stepping aside would be best for Plant Genetics. In a letter to Henry and Nickell, I explained that while creating a company to clone third-world perennials was important, I would be declining the senior scientist position. It had become obvious that my values and approaches to people were in serious conflict with Henry's.

*****

I took a few long runs to refocus my mind and returned to preparing for our late fall departure to Davis. I shared the biology secretary in Rensselaer's Cogswell Laboratory with two full professors. As the fall semester began, I requested from them first access to her time until I left for Davis. For three months, I sat at my lab desk from morning into night, writing or editing and guiding a technician, two senior thesis students, and Singer. I was also teaching the first half of Introduction to Organisms, a lecture and lab with 84 students. There was much to do. The renewal grant for NIH funding had scored too low for support, so I appealed the decision with letters and a revised proposal. I wrote two manuscripts on amino transport, an article with Singer on amitole transport, a review with Singer on mutant selection, and a techniques and methods chapter on the transport of ions and organic molecules. My two senior colleagues were not happy; for 10 weeks I used 100 percent of our secretary's time as she typed the penultimate and final copies of what I had written.

In the midst of all this writing, I received an invitation from Dennis Carr in Australia and Peter Barlow in Great Britain to write a chapter for a book titled *Positional Controls in Plant Development,* to be published by Cambridge University Press. The idea had emerged in August during the International Botanical Congress in Sydney and included a roster of possible contributors from around the world. With delight, I accepted, knowing that my developmental perspective and papers were gaining international standing among plant scientists with stature in the field. Due in August 1982, the manuscript would become a substantial part of my sabbatical at Davis, which was the perfect location to write, given its excellent library resources and plant science colleagues.

On the home front, Mary had resigned her position as teacher and director of a preschool and was preparing us for a pre-Thanksgiving departure. She packed and mailed boxes of things we would need in the house we were renting. My friend and colleague Tom Barnett, who had joined the faculty at State University New York-Albany, would live in our Troy house. Barnett had become a part of our family beginning with our days together in Berry's lab at Wesleyan. While a postdoc at Brandeis University, he had helped paint our house in summer of 1979, a rather boring task. To pass the time, we talked about all manner of things, including, of course, science. We considered heat-shock proteins that had been discovered recently in animals, but not observed in plants. We now know that heat-shock proteins are ubiquitous in all cells, from microbes to eukaryotes—organisms whose cells have a nucleus—and are chaperones that facilitate the normal folding of proteins. In cells stressed by heat or other conditions, they ensure proper folding of damaged proteins. On a lark, we decided to look for them in cultured plant cells.

I had heat-shocked suspension-cultured tobacco cells and then sent an extract from homogenised cells to Barnett at Brandeis for gel electrophoresis to establish if the heat-shock proteins were synthesised. Cultured tobacco cells expressed the classic heat-shock proteins observed in fruit flies, a clear indication that they had evolved in an ancient ancestor of animals and plants. As it turned out, another colleague, Joseph Mascarenhas, a professor at SUNY-Albany, had a graduate student observing cultured soybean cells to see if they expressed heat-shock proteins. They confirmed and added to our observations. We pooled our efforts and in 1980 published the first report on heat-shock proteins in plants in *Developmental Genetics*. Barnett was listed as first author, because he was job hunting.

On Saturday, November 21, 1981, my family and I headed southwest. On the way, we stopped in Washington, D.C., to meet with Holly Schauer, associate director of the USDA competitive grants program that was funding our laboratory's herbicide resistance research. Along the way, I also gave seminars at the University of Tennessee and Texas A&M. After a month, we arrived in Davis, early enough for Stuart and Virginia to attend two days in their new schools and to spend Christmas with Mary's aunt and uncle in Orinda.

I sought out Hackett and Sachs. We hit it off well, and they asked me to participate in the department's graduate seminar. I engaged the seminar participants in constructing my chapter in *Positional Controls in Plant Development*. Whenever I found a paper that inspired me, I would bring it to the graduate seminar for analysis and discussion. Sachs apparently appreciated my involvement, as he wrote an affirming letter to Joe Landau at Rensselaer. Writing the chapter, along with reading and talking with faculty and students in the Department of Environmental Horticulture for hours, became the major accomplishments of my sabbatical. When back at Rensselaer, I sent the penultimate draft of *Positional Controls* to Sachs for his final input. He wrote, "I like the paper—as is. It reflects a point of view not expressed in the botanical literature. Particularly for plant physiologists, the chapter will be very stimulating."

We headed home in early June, car-camping as we wandered east via Mono Lake, the Grand Canyon, Milwaukee to see Grandma McDaniel, and on to Troy. The Grand Canyon was the highlight of that trip. Mary and I had attempted to reach the south rim in December 1967 after our wedding but were turned back by a snowstorm. This would be our first time visiting this breath-taking natural wonder. We would later enjoy two six-day backpacking adventures there.

We stayed overnight at Phantom Ranch, just a bit north of the Colorado River where the trails from the north rim and south rim meet. Virginia was eight and Stuart 11, but both were able to hike down the seven-mile South Kiabab Trail (a 3,500-foot descent) and up the 10-mile Bright Angel Trail (a 3,000-foot ascent). The June heat only added to the experience. When we reached the Colorado River, the temperature was about 105 degrees. We took off most of our clothes and jumped in, but for a short time—the water temperature was perhaps 40 degrees. The sand was hot on our feet, so we jumped in and out of the cold water. After tiring of the extremes, we headed to Phantom Ranch.

Grand Canyon National Park is 277 miles long, up to 18 miles wide, and over a mile deep. The Colorado River has been cutting into rock for perhaps five or six million years as the land around it rises, although recent uranium-lead dating indicates a much longer making of the canyon. The geology provides a spectacular layered sequence of more than 36 major layers of exposed sedimentary rock, ranging in age from the 200-million-year-old Kaibab Limestone on the southern rim to the 2-billion-year-old granite at the river. These sedimentary rocks hold a treasure trove of fossils representing two billion years of life. I took photos for slides while on our hike; and for two decades used them in my introductory biology courses.

The IRS audited our tax filing that year because our deductions nearly equalled our modest income. I met with the head of the Albany IRS office; she commented, "You cannot deduct your Grand Canyon vacation expenses." My response was most enjoyable for me. After explaining the geological and biological importance of the canyon, I presented her my 3-inch-thick binder of the 500 slides I used in the classroom. In the end, the IRS owed us a little over 100 dollars.

*****

At the end of the summer, coincident with sending to Barlow my chapter for *Positional Controls in Plant Development*, I received from George Malacinski, series editor of *Primers in Developmental Biology*, a request to write an introductory review article for *Pattern Formation: A Primer in Developmental Biology* on a topic of my choice. Twenty-six developmental biologists from around the world wrote on organisms ranging from ciliate protozoa and *Volvox* to higher plants and animals. Three chapters focused on plants: *Controls and Cell Patterns in Plants* by Tsvi Sachs, *Patterns and Problems in Angiosperm Leaf Morphogenesis* by Scott Poethig, and my chapter, *Competence, Determination and Induction in Plant Development.* Members of the 1970s Sussex group had written all of the higher plant perspectives in this seminal book on pattern formation.

Although the animal developmental viewpoint I was presenting to botanists and plant physiologists was relatively new to them, I was doing it with day-neutral tobacco in the genus *Nicotiana.* This genus had other species and varieties whose flowering was mitigated by photoperiod. Especially Anton Lang in the U.S. and Makhail Chailakhyan in the Soviet Union had extensively employed photoperiodic *Nicotiana* in flowering studies. Both had been esteemed scientists in flowering research for half a century. I never met Chailakhyan, although I read a number of his papers and extended some of his conclusions.

4.3: Image of Anton Lang.

Many of my colleagues who studied plants would not submit manuscripts to *Planta* because Lang was the editor. They considered Lang a prickly, biased editor who carefully reviewed and meticulously edited each submission. I suspected Lang would be an even higher hurdle for me because my node-counting studies fell in his area of expertise: flowering. Nevertheless, in October 1979 I sent *Planta* a paper titled, *Influence of leaves and roots on meristem development in Nicotiana tabacum cv. Wisconsin 38.* Lang's approval would authenticate its quality.

His typed, three-page, single-spaced reply came late in December. It included my manuscript, which he had hand-edited from title page through references: "does your paper really show this?"; "could be said more summary-like"; "this is hard to follow"; "meaning?"; "this essential?"; "only people can suggest"; and "a repeat". Lang had done what Scott did 18 years ago to my literature review on auxins and gibberellins—hacked my submission to pieces. Additionally, Lang's critiquing letter was half the length of the paper itself, totalling six pages when double-spaced.

Lang wrote:

As to the factual data, you demonstrate very clearly that roots play a major role in the regulation of the transition of the shoot apical meristem of a dayneutral tobacco from vegetative growth to flower formation, proximity of roots to the shoot apex evidently being inhibitory to flower formation, while leaves have no such effects. These results are certainly worth communicating. The only reservation here is that this is done at considerable length.

He then explained how to eliminate one of five tables and five of six figures and how to shorten the text by some 40 percent. In February, I sent the revised manuscript. Lang quickly replied:

Your revised paper has arrived and has been sent on to Springer Verlag. Using your permission, I have made a few minor changes, some to compact the writing still a little more… and one or two more to make a statement more precise… As far as contents and presentation, the paper is now entirely fine for *Planta*, and the results are of considerable interest.

I had had my first lesson from the person who would, more than any other, enable me to strengthen my writing. Additionally, these two years of experimental results had passed muster with the world's leading expert on flowering in *Nicotiana*. To offer a flavour of the science presented, let me describe the key experiments and their results. Decapitation of a plant above the fourth or 10th bud below the inflorescence resulted in the bud growing to flower with a precise number of nodes, few for the fourth bud and more for the 10th bud. Leaving no leaves or four at locations below the bud did not change the number of nodes produced before flowering. These results indicate that leaf position and number do not influence the number of nodes produced before terminal flower formation.

We induced roots on the stem by dropping a series of bottomless pots around a plant as the shoot grew, filling the pots with soil so as to stimulate root formation on the stem. These pot-on-pot plants produced 10 or more nodes than did the untreated control plants. As an additional control leaves were removed from a group of plants at the same rate as stems and leaves of pot-on-pot plants were buried in soil. These plants produced the same number of nodes as untreated plants.

4.4: Pot-on-pot plants.

In a second procedure, two groups of plants had the upper portion of the stem rooted each time the plant had six to 10 leaves of a certain size—one group rooted 10 times, the other 14 times. Both groups were then allowed to grow and flower. This produced more than 100 nodes for group one and more than 130 for group two. I asked Lang how long tobacco flowering would need to be prevented before assuming that the plant would never flower. "One year," he said. Our re-rooted plants were prevented from flowering for well over a year. These two rooting experiments established that roots close to the terminal shoot apical meristem prevent the plant from flowering.

By this time my laboratory's investigations of node counting by *N. tabacum*, we knew four things about the growth of tobacco shoot apical meristems. First, the number of nodes produced by an axillary bud meristem is position-dependent; the higher on the stem, the fewer nodes produced before formation of the terminal flower. Second, the first, second, and third axillary buds below the floral branches may be determined for a few-node growth pattern with a one-internode separation between determined and undetermined buds. Third, only a few leaves at any location are needed for a shoot apical meristem to flower after producing the expected number on nodes. Fourth, roots relatively close to a growing shoot apical meristem inhibit the transition from vegetative to floral development.

I myself did the experiments that enabled the above understandings; aside from the help I had growing plants at Yale. This was about to change. Although we applied to the NIH, NSF, and USDA for extensions of our amino acid and herbicide tolerance projects, none were successful. On the positive side, three new graduate students were interested in determination; that, combined with the NSF funding for growth pattern analysis, shifted the emphasis of our lab.

In 1985, Kelly Dennin's master's thesis on determination for growth-pattern in axillary buds of the long-day plant *Nicotiana sylvestris*—plants that flower only when grown in a long-day photoperiod (e.g., 16 hours of light, eight hours of darkness)—yielded a most interesting observation. The experiments were done under long-day conditions with plants at the opening of the terminal flower. To start, Dennin showed, all in long days, that *N. sylvestris* plants count nodes precisely, 36 nodes. She established, similar to *N. tabacum*, that if the shoots were continually rooted, they would not flower until the plants were allowed to grow without being re-rooted. Additionally, some of the first few axillary buds below the floral branches were determined for a limited growth pattern as assayed by rooting and grafting of buds to the base of another plant. Similar to *N. tabacum*, one internode separated determined from undetermined buds. However, determined buds that were rooted did not produce normal inflorescences. Instead they produced a single flower along with one or two vegetative stems that produced many nodes before forming normal inflorescences. These results appeared to indicate that the whole meristem was not stably determined for a limited growth pattern; rather, only a population of cells in the centre of the meristem remained in a florally determined state in the presence of nearby roots. Simply put, a population of cells in the centre of the meristem is determined to make a flower while neighbouring cells were either not determined or reverted to vegetative growth.

4.5: Nicotiana sylvestris rooted axillary bud with abnormal inflorescence.

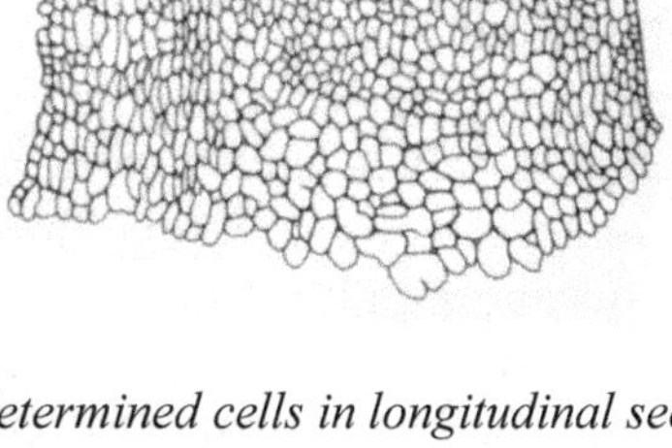

4.6: Diagram of florally determined cells in longitudinal section of meristem of rooted Nicotiana sylvestris axillary bud.

I was excited to have an example of cells that were apparently determined to form a flower. For years, it had been known that tissue explants from floral branches of *N. tabacum* regenerated new shoot apical meristems that either formed a small plant with few nodes and a terminal flower—floral shoots—or a vegetative shoot with many nodes. These floral shoots exhibited a gradient with a high percent of floral shoots from floral branch tissues with the percent dropping to zero as tissues were taken further down the stem. I had discussed these observations in my first *Developmental Biology* publication, in which I reported that a one-internode separation existed between determined and undetermined axillary buds. In that paper, I hypothesised that this internode on a given plant was also the boundary between internode tissues that formed floral shoots and those that did not produce floral shoots.

By the time Singer was finishing her master's thesis, she had decided to devote her PhD thesis to testing my hypothesis on the coincidence of the boundaries for florally determined buds and floral shoots regenerated from cultured tissues. In her senior thesis and master's research, she had demonstrated her capacity to take charge and independently navigate her way through complex, challenging tasks. This had resulted in six primary research publications, a rare accomplishment for a master's degree. Singer had become a colleague whose PhD thesis would demonstrate in four independent ways that my hypothesis could not be supported. Let me describe what her thesis established.

Initially, we believed her findings should be presented in a single paper because we believed it to be a single coherent unit. We pursued publishing in the *Proceedings of the National Academy of Sciences (PNAS)*, because we felt our experiments were exceptionally well done and that the conclusions advanced substantially our understanding of determination in tobacco. Only a member of the National Academy of Sciences can submit a paper for publication in *PNAS*, so I sent a letter and summary of our results to Lang, asking if he would consider doing so. He replied positively, stating that two reviews are required and asked us to suggest two or three who could "evaluate this work knowledgeably and whose judgment you would respect. I would try and get them to review the paper." He suggested two; we suggested three more. Three people sent reviews. One recommended the paper's publication if two minor issues were addressed. The other two reviewers and Lang expressed more concerns. Clearly, we had presented the results poorly. From the comments, we recognised the benefit in splitting the results into two papers: one on the timing of determination in the terminal bud and axillary buds, and the other on the temporal and spatial appearance of floral shoots in cultured internode sections. We sent the first to *Developmental Biology*, the second back to *PNAS*.

We employed histological procedures and our rooting assay for determination to show in the *Developmental Biology* paper that 1) the terminal bud became florally determined before the terminal meristem initiated the gross morphological changes associated with floral initiation; 2) the terminal bud went from being in a vegetative state to being florally determined in about two days; 3) after being florally determined, the terminal meristem made about three more nodes before forming the terminal flower; 4) axillary buds shown to be florally determined were produced from cells derived from the terminal apical meristem prior to the terminal bud becoming florally determined; 5) axillary buds shown to be florally determined were formed about five days after the terminal bud became florally determined.

These results can be summarised as follows. The growing terminal bud became florally determined over about two days prior to the morphological initiation of a floral meristem. Floral determination was thus not clonally inherited by axillary buds from the florally determined terminal bud meristem; rather, a pervasive signal appeared to act independently on the terminal and axillary apical meristems or on the groups of cells that form florally determined axillary buds.

The second paper appeared a year later in *PNAS*. We established two things about florally determined shoots generated by cultured internode tissues. First, internode tissues gain the capacity to form floral shoots in culture at about the same time that the terminal bud becomes florally determined. Second, the first internode tissues to form floral shoots in culture were located about 24 nodes below the terminal meristem (about 15 centimetres). The concluding paragraph summarised why the paper was important.

The fact that floral determination occurred synchronously in the terminal bud and in spatially distinct internode tissues has general implications for the fields of flowering and developmental biology. Florally determined internode tissues below the inflorescence are not clonally derived from florally determined apical meristem cells; rather a common signal(s) appears to act on shoot apical meristems, as well as on internode tissues, to bring about a new developmental state, floral determination. Cells need not be organised as shoot apical meristems in order to acquire this developmental state. Acquisition and expression of floral determination are distinct phases of floral development and can be separated by many days. Physical parameters of the meristem may be important in the actual formation of flower parts; however, they cannot explain floral determination in internode tissues [Researchers had hypothesised that physical forces in a shoot meristem might influence determination.]. The acquisition of this state appears to be an inductive process in which a developmental signal acts on competent cells to bring about a specific state of determination.

Several of the statements in this paragraph could have been made about determination in animal development for decades; however, not all were well documented for plants. For the field of flowering, we clarified the temporal and spatial aspects of the gradient of floral shoots produced by cultured internode pieces in day-neutral tobacco and established more explicitly what we meant by floral determination.

The story that led to the publication of the *PNAS* paper is notable. As stated above, our first submission led to the publication of two papers. The part of the original *PNAS* manuscript that we sent back to *PNAS* was revised four times and reviewed five times in 1986. Lang wrote 10 pages of comments in eight reviews; three reviewers wrote 11 pages of comments in nine reviews; and Singer and I wrote nine pages of explanations for the issues raised by Lang and reviewers. We were exhausted from the yearlong process. After the fourth revision, I sent this statement to Lang: "Well, there it is. If every manuscript went through what this manuscript has gone through, we would have about a dozen journals in this country and every article would be about as close to excellent as the standard person is capable of doing." Lang's response, "Well, Carl, if every manuscript went through what yours has gone through, I would say there would not be any scientist left wanting ever to write a paper, and no reviewers and editors wanting to review or edit."

*****

Seven years had passed since my semester at UC-Davis, and I was again eligible for sabbatical. I had been teaching an embryology course off and on since first arriving at Rensselaer. The lecture portion of the class focused on the developmental biology of multicellular organisms, in which I presented the classic procedures of fate mapping and clonal analysis. Fate mapping of animal eggs and embryos had long been possible because an area or a cell was accessible to being marked with dyes. The progeny of marked cells in an area, or of a single marked cell itself, is a clone and can be identified at a later stage of development because it carries the dye. If a group of cells in an arm limb bud is marked and the progeny of these cells form a thumb, we know the fate of the marked cells is a thumb. In this way by analysing the fates of all cells in an arm limb bud a fate map of the entire arm limb bud can be made.

Plant eggs and embryos are embedded in a seed and cannot be marked with dyes or by other physical means. However, in the 1960s it became possible to genetically mark a plant cell by creating a mutation in pigment synthesis at a particular stage of development and to observe the area covered by the daughter cells of a pigment-marked cell later in development. The clones of pigment-marked cells are analysed to produce the fate map. However, like so many things, I did not understand in depth fate mapping by clonal analysis until I did it.

Scott Poethig, my colleague from the Sussex lab who was now teaching at the University of Pennsylvania, had done clonal analyses of leaf formation in tobacco and was interested in creating a fate map of the corn shoot apical meristem at the dormant embryo stage. Similar fate maps of the corn meristem had been interpreted by some researchers to indicate that the tassel, the male reproductive part of the plant, was produced from a group of cells isolated in what animal developmental biologists called a developmental compartment. A compartment was indicated when a distinct morphological boundary existed that marked cells did not cross in the final structure they formed; for example, clones derived from marked cells in a wing of the fruit fly did not extend into the adjacent thorax because the wing is formed by cells in a wing disc that just forms the wing. The evidence for a tassel compartment was that clones derived from marked cells in the tassel did not run across the boundary between tassel and the rest of the plant below. The compartment concept had come from fate mapping in animals where it made sense to Poethig and me, whereas it did not in flowering plants. I spent the summer and fall of 1987 on sabbatical in Poethig's lab to test the tassel compartment hypothesis.

Under Poethig's guidance, we made a fate map of the corn embryo apical meristem that turned out to be the most comprehensive one done at the time. With gamma radiation, we created anthocyanin pigmentation mutations that were green and sun-red, making them relatively easy to spot in dark purple-coloured corn plants. I spent most of August and September sitting at a card table in Poethig's cornfield recording positions and extents of 277 clones in some 160 plants out of the 1,050 that matured after radiation to create pigment mutations. Being out in the open on warm, sunny days among the sounds of nature made tedious work tolerable, if not enjoyable.

During my sabbatical, I would spend two weeks in Philadelphia, then return home by train for several days to keep the research going smoothly and have time with my family. In reality, while my time at Davis was truly a sabbatical, this was more like holding two full-time jobs—the price one pays for the freedom of being one's own boss in a profession that demands more than can be delivered and provides rewards of accomplishment among kindred spirits seeking knowledge and understanding instead of economic gain.

After the fieldwork, I spent the next three months analysing data and writing a paper for *Planta*. Since our intense experience of making the *PNAS* manuscript meet the high bar of Lang and his reviewers, I had adopted a new procedure before submitting a paper. I would write the best presentation possible and select two of the most critical scientists I knew in that topic to pre-review it. In this case, I chose Ed Coe Jr. from the University of Missouri-Columbia and Michael Christianson at the Zoecon Corporation.

Anton Lang, as usual, picked excellent reviewers: himself and three others, who sent seven pages of comments for us to address. The reviewers recommended publication with revisions. We revised in accordance with the comments; and the manuscript was accepted. The pre-reviews apparently resulted in a better manuscript with fewer stumbling blocks.

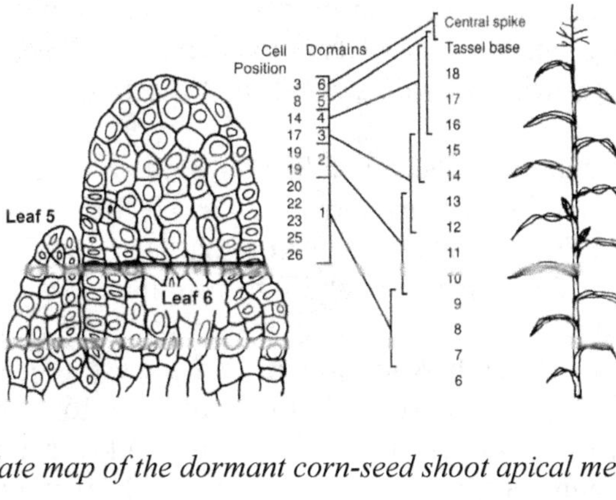

4.7: Fate map of the dormant corn-seed shoot apical meristem.

Our data and analysis showed that the tassel was not a compartment, nor did any cell in the embryonic corn shoot apical meristem have a predictable fate in the adult. In addition, we explained how earlier fate maps could have made a tassel compartment a legitimate conclusion. Although Lang and one reviewer were not impressed, this paper was studied in many courses and graduate seminars. It was also the first paper of mine to be cited more than 100 times in the primary literature, a major accomplishment. Poethig told our son, also a plant scientist, that the work I did in his lab was instrumental in his earning tenure.

*****

As our flowering work gained recognition among the plant and developmental biology communities, I received invitations to present our studies to international scientific gatherings. In England in 1987, I spoke at the 45th Easter School in Agriculture Science titled *The Manipulation of Flowering.* There I met Lloyd Evans, a past president of the Australian National Academy of Sciences and, like Lang, one of the giants in flowering research. I considered finding time to go to Australia to develop a procedure to culture apices of *Lolium temulentum*, Evans's flowering workhorse. Later that year, at the UCLA Symposium on Plant Gene Systems and Their Biology, I crossed paths with Jim Peacock, head of plant industry at the Commonwealth Scientific Industrial and Research Organisation (CSIRO) in Canberra where Evans also worked. Peacock was a molecular biologist investigating flowering in various plants, including tobacco, and asked if I would come and interact with CSIRO scientists. He commented, "I will see if I can offer a return trip for you." I responded, "So I will pay for my flight to Australia?" He laughed and explained that in Australian English, return trip means round trip. I was to find out these two English languages had many divergences.

In June the next year, I received a formal invitation from Peacock to visit and work at CSIRO. The job would provide a living allowance and an airline ticket on QANTAS (Queensland and Northern Territory Aerial Services). I agreed to a three-month visit beginning in spring 1989 and began coordinating with Lloyd Evans and his colleague, Rod King, with whom I would be working, the processes of getting tobacco seeds cleared to enter Australia and arranging for *L. temulentum* to be growing when I arrived. In November, a letter arrived from King stating,

Congratulations on your success in being awarded a McMaster Fellowship. Maybe this is the first time you have received an award without applying for it. Anyhow, I enclose the final pages of fiction that led to your success and hope you can live with it. Was it the Scottish lilt to your name that attracted them!?

This was quite an honour and totally unexpected. Sir Frederick McMaster was a 'prominent New South Wales grazier' whose gift to CSIRO led to three fellowships annually in all of Australia to be awarded to "support very senior and distinguished overseas scientists selected to work for a period in CSIRO Divisions." The nomination came from Peacock, and King wrote the application.

The award was quite generous: two round-trip business class air tickets ($9,600), funds for internal travel to give seminars ($2,128A), a living allowance ($5,000A), and research funds ($2,700A).

I asked King to send me the money for the airline tickets instead of the business class tickets themselves. This would allow me to purchase five coach tickets; three for the family and two for me. I knew I needed a brief trip back to the U.S. in June for a meeting on root development, an editorial board meeting for *Developmental Biology*, and the advantage of being able to attend our son's high school graduation. My request for the cash was unusual—CSIRO had no guarantee that I would actually show up—but King made it happen. Trust is the quality that allows so much to be accomplished.

Stuart, our son, was not particularly happy about traveling with his parents, so we arranged for his best friend, Chuck Borton, to travel with him for the summer in Australia. We gave Stuart airline tickets, $2,000, and three days on the Great Barrier Reef on the condition he would come through Canberra on his way home. Stuart and Chuck had most memorable precollege adventures.

In late June, I met the family in Christchurch, New Zealand, for a short holiday. During dinner the first night, Stuart asked, "Dad, just in case someone asks, what are you doing here?"

I started answering, "I'm studying flowering in *Lolium temulentum*—"

Chuck interrupted, asking, "Like in Linoleum kitchentum?" We all laughed. That was the end of the explanation.

My father had always wanted to see Australia, but never did. Here I was, traveling with my family, essentially all expenses paid by the Australian government. We had a splendid first time 'Down Under'. Nevertheless, for the first three months, I worked 24/7, and when the family came, as intensely as I worked at Rensselaer. I lived in a guest dorm at Australian National University, a five-minute walk from the Phytotron, which was a complex of greenhouses, growth rooms, and cabinets that provided state-of-the-art programmable growth conditions. Evans, King, and other plant scientists had offices and other support facilities connected to the Phytotron. At the time, Australia, a country of about 20 million, had many internationally known and accomplished scientists, in part because CSIRO hired the best young scientists from around the globe, gave them what they needed, and then turned them loose to collaborate among leading scientists everywhere. To be honest, I wondered why I was there.

I had two major projects underway at the Phytotron. Under various precisely controlled conditions, we studied the number of nodes produced by several varieties of tobacco that produced few nodes. I worked with a CSIRO scientist on this project, but in the end, nothing publishable came of it. The second project consumed most of my time. My goal was to culture apices of *L. temulentum*, a long-day monocot that flowers with exposure to a single long-day, to establish when an isolated apex after one long day would be florally determined, defined as when an isolated apex became capable of initiating floral morphogenesis. To do this, I needed to devise a culture medium in which excised apices would survive and at least initiate floral differentiation, if not make normal flowers. I expected to bring the *L. temulentum* culture procedure to Rensselaer for further exploration of floral determination, a developmental process. Evans and King, as plant physiologists, were interested in culturing apices to explore sugar uptake, abscisic acid (a plant hormone) levels after excision, the influence of gibberellin (another plant hormone) on cultured apices of short-day grown plants, and floral differentiation in culture compared to that of intact plants.

I was pretty sure that had Evans and King been reviewers of my early floral determination grant proposals and tobacco papers, their reviews would have been negative, as they lacked a developmental perspective. Before arriving in Canberra, I had written down for discussion with Evans and King what our eight *N. tabacum* papers had contributed to our understanding of flowering. Once in Canberra, we spent tens of hours talking, on many subjects. We scheduled time to discuss each other's publications, where I often articulated a developmental perspective. Evans had been studying the physiology of flowering in *L. temulentum* since 1960 and gave me the take-home messages from his publications relevant to flowering. He used the term 'evocation' to describe the induction process at the apex caused by the floral stimulus from the leaves—the hormone proposed in the 1930s by Lang and Chalaykian that promotes flowering and that was yet to be identified. In early discussions, he had not seen a need to introduce floral determination into the flowering literature. After many hours of discussion, I believed he accepted my developmental perspective as legitimate and useful. In my mind, we had arrived at consensus when agreeing that the precise defining of the time of floral determination in *L. temulentum* would establish for the first time in any plant the maximum time required for the process of floral evocation that committed a meristem to form flower(s).

Evans and King were gracious hosts, lending me a bicycle and helping me to acclimate to the Phytotron and adjust to life in Australia and its customs. When Mary and Virginia arrived in July, Evans kindly invited us to stay at his house until our September departure, as his wife was working overseas. Evans and King had us to their beach houses. They made our time in Australia so satisfying that Mary and I returned three more times.

We met the apex culture challenge in three months and began doing experiments. In late August, we had the data indicating when cultured apices were florally determined, but Evans and King had to repeat critical experiments after I left. Conversations with them no longer had rough edges concerning my developmental perspective. We considered where to send our paper, perhaps attempting a publication in *Nature* or *Science* that could be expanded in *Developmental Biology*, the top-ranked developmental journal in the world at the time. I served on the editorial board and was seeking more plant papers. A month or so after returning to Troy, our exchanges over the emphasis and language used in the manuscript mirrored the physiological and developmental perspective differences that had animated our early discussions in Australia. To me, it appeared as though we had never arrived at a consensus. I wanted 'determination' in the title; Evans and King did not. They wanted to publish the paper in *Annals of Botany* (a fine but ordinary plant journal), not *Developmental Biology*. Since my departure, they primarily used the culture of apices to examine the influence of other compounds on the floral differentiation of apices—the effect of sucrose concentrations with and without mannitol (another sugar), kinetin and zeatin (two plant hormones), and other compounds—and wanted the paper to emphasise those physiological data. Over the next year, we exchanged dozens of letters and faxes, attempting to resolve our differences.

Throughout this controversy, I talked often with my father-in-law, Donald Frazier. I had discussed innumerable situations with him over the years, often seeking his opinion. After months of observing my contentious interactions with my colleagues, he told me a story from his graduate school days in political science. He was in a course with small groups that had an assigned topic for which they were to find consensus. His group consisted of a Jew, a Christian fundamentalist, and a Muslim; he was a mainstream Protestant. The topic was to determine if women should be allowed to vote in political elections. Despite their best efforts and hours of discussion, they could not reach consensus. Frazier concluded that I would be unable to resolve my differences with Evans and King enough to publish a joint paper. Despite his scepticism, 19 months after leaving Australia, our co-authored paper titled, *Floral Determination and in vitro floral differentiation in isolated shoot apices of Lolium temulentum L.* was accepted for publication in *Planta*.

With support from my Australian colleagues, we brought *L. temulentum* to our lab and began using the culture procedure to further explore flowering. Our work in Australia and setting up *L. temulentum* apex culture in my lab became major factors in the NSF and USDA awarding funding for a two-year proposal entitled, *Developmental States Associated with Flowering.* We could accept just one grant for the same proposal, so we took the larger USDA grant. This enabled our lab to support two undergraduate theses on flowering in *L. temulentum* (one of which won the 1902 Prize for Best Institute Undergraduate Thesis), a technician, a third PhD student, and a postdoctoral fellow. I published with my technician a paper that indicated two sequential signals regulated floral initiation in *L. temulentum*. The long-day-stimulated-leaf-signal put the apex in a florally determined state that is then elicited by gibberellin. The title of the paper was, *Flowering as metamorphosis: two sequential signals regulate floral initiation in Lolium temulentum,* which hearkened back to my insect research in which three signals were involved in regulating insect metamorphosis.

I published several more papers on flowering before closing my research lab to take on the role of founding director of Rensselaer's undergraduate environmental science degree program. It is appropriate to close this chapter with what more than two decades of research told us about node counting.

In the early 1980s, I realised that I could use tobacco plants that produced a wide range of nodes to investigate the genetics and developmental physiology of node counting. I found the genetic diversity I needed at USDA's Tobacco Research Laboratory. It housed the USDA tobacco germplasm collection that had over 1,000 genotypes called Tobacco Introductions (TIs). We selected 35 that made few nodes (15 to 30) and 37 that made many nodes (38 to 68). From each group we identified the best node counters for genetic study and for grafting of various TI shoots onto various TI roots.

Ten years of work by two technicians, four undergraduates, three graduate students, and me yielded a paper titled, *Regulation of node number in day-neutral Nicotiana tabacum: a factor in plant size*, published in *The Plant Journal* in 1996. The summary from the journal article read:

Employing genotypes of day-neutral tobacco that exhibit a wide range in the number of nodes produced, it has been established that node number, and plant size, in tobacco are regulated, in large part, by two endogenous signals and one developmental state, competence. All genotypes have the same level of a root signal that maintains a vegetative pattern during early growth. The number of nodes produced before the formation of the terminal flower, as well as plant size, is a function of the strength of the floral stimulus from the leaves and the competence of the terminal meristem to respond to the floral stimulus by initiating the terminal flower.

As this paper illustrates, it often takes an experimental scientist many publications over decades of time to assemble the information necessary to answer an easy to state fundamental question. In this case, "How does the tobacco plant count the number of nodes produced before ending growth by forming a terminal flower?" After all, the goal of scientific research is to tell a story that is a true story, told with the data and the analyses of that data that are consistent with all the relevant stories that have come before.

Chapter 5
Professional Service

Whom do you know? Whom do you trust? The answers are foundational for social relations in all human affairs, be they jobs or organisations in business, medicine, politics, science, or sports—or even in groups and clubs ranging from historical societies and religious assemblages to street gangs. For the most part, people self-aggregate into particular associations via cultural guidance, but in an open society also by choice. Scientists, like everyone else, form relationships that shape a trajectory of success or failure as defined by their accomplishments and stature among peers.

The Holy Grail of science is discovery, as noted in my previously mentioned quote from Wilson: to be a real scientist is to make a certifiable scientific discovery affirmed by scientists in the same area of expertise. In publications, scientists reveal discoveries that have been authenticated by other scientists who are deemed qualified to assess the validity of the findings. The resulting paper is then put on display in a scientific journal for scientists to examine. If something of great importance is presented, some scientists will attempt to repeat the work, while others will authenticate its veracity by its predictive power or correlation with related phenomena. In time, an initial hypothesis that explains the observation is supported or refuted. A terse retelling of the discoveries that concluded that deoxyribonucleic acid (DNA) harbours genetic information for essentially all life on earth illustrates how scientific exploration attempts to uncover truth.

Today, all biologists accept DNA as the molecule that transports the information for making life from inorganic material. If in 1920, however, you had claimed that DNA was the molecule containing information for making an organism, you would have been considered wrong. The presumed answer was protein. At that time, we had no direct evidence that any molecule carried genetic information; however, of the choices—carbohydrates, DNA, fats, proteins, ribonucleic acids (RNA)—only proteins appeared to have sufficient complexity to encode genetic information. That is, they are large molecules of a linear sequence composed of some 20 distinct amino acids yielding an immense diversity of sequences. DNA and RNA are each composed of just four subunits (nucleotides), fats are mostly long chains of carbon atoms bonded to each other and with two hydrogen atoms, and carbohydrates are linear and branched chains of sugar subunits (e.g., glucose, fructose, and others depending on the specific carbohydrate).

Two sets of experiments established that in one bacterial species (*Streptococcus pneumoniae*), DNA was the source of genetic material. In 1928, while studying pneumonia in mice, British microbiologist Frederick Griffith showed that a mutant strain of *S. pneumoniae*—one that didn't cause pneumonia—could be changed into the virulent form if the heat-killed strain of virulent bacteria was injected into mice that had non-virulent bacteria growing in them. The process by which this virulence was acquired is called transformation. The quest was on for the chemical structure of the transforming principle.

Fourteen years later, scientists Oswald Avery, Colin MacLeod, and Maclyn McCarty performed a set of experiments identifying DNA as the transforming molecule. And while those experiments today are considered the definitive proof that DNA is the genetic material in bacteria, biologists at the time were not convinced; the DNA that apparently effected transformation was not 100 percent pure—it contained protein at 0.002 percent. For decades, biologists had believed that proteins were the genetic material—a difficult belief to abandon.

In 1952, Alfred Hershey and Martha Chase offered the convincing evidence that DNA was, in fact, the genetic material in bacteria. Their experiments established with a bacteriophage (virus of bacteria) that virus DNA, not protein, carried viral genetic information. In 1953, James Watson and Francis Crick, with the help of Rosalind Franklin—whose X-ray diffraction patterns showed a helical form for DNA—established the structure of DNA, which provided a model for replicating with fidelity the nucleotide base sequence in a DNA molecule. All biologists assumed, without any direct evidence, that DNA was not only the genetic information for bacteria, but also for animals and plants. Direct evidence came in the early 1970s, when recombinant DNA technology permitted moving pieces of DNA from one organism to another. That is, DNA from bacteria could be inserted into plant or animal DNA and be expressed, and vice versa.

This story illustrates two essential aspects of science that are present to a lesser degree in other human activities: science is both a community endeavour and an accretive process. Each of the discoveries mentioned earlier, including Griffith's, emerged from groups of researchers. Furthermore, Griffith needed earlier studies on pneumonia in mice to make his own discovery. Avery, MacLeod, and McCarty then built on Griffith's insight. Hershey and Chase added on to preceding studies, as did Crick, Franklin, and Watson. Scientific discovery is an ever-expanding starburst of established knowledge with myriad links across the display that is forever revised with new information and modifications of existing understandings. Moving forward, no one needs to rediscover the nature of the transforming principle or re-identify the genetic molecule in plants unless someone finds that earlier studies missed something or made a flawed assumption that questions established knowledge. All scientists seek to prove important, accepted knowledge wrong or discover something of major importance. But these are high bars cleared by few scientists!

The stories that relay the discovery of DNA as the carrier of genetic information are taught in introductory biology courses everywhere and are indicative of all science. Every scientific discovery depends upon an infrastructure that requires scientists to participate in professional alliances and activities. Examples of this professional service infrastructure were outlined in the last chapter: reviewing manuscripts, serving on faculty search committees, writing letters of reference, supervising student research, organising conferences, and publishing on scientific topics.

*****

Michael Freeling, a corn geneticist at the University of California-Berkeley and later a member of the National Academy of Sciences, invited me in 1983 to speak at the symposium 'Competence, Determination, and Canalisation in Plant Development', part of a joint meeting of the Botanical Society of America and the Canadian Botanical Association. He had likely heard about my work from a colleague or read my publications. In any event, he thought my research important enough to be heard at an international gathering of botanists. Freeling was providing a service to the botanical community, while I was given a platform on which to present my science for approval. What follows are two stories of a symposium and a meeting that I organised and co-organised, respectively.

Plant Physiology is a journal well regarded by plant scientists. Our lab published six papers on amino acid transport and herbicide resistance in the journal, but none on our flowering work because developmental biologists did not read it. In 1990, due to my growing reputation for advocating a developmental perspective in plant research, I was invited to the program committee meeting of the American Society of Plant Physiologists (ASPP, renamed in 2001 as the American Society for Plant Biologists) that would pick symposia topics for its 1991 annual meeting. I suggested two topics but declined to be a speaker so as not to be distracted from effectively organising and running the event.

The program committee chose my topic, *The Genetic Analyses of Plant Developmental Processes* and provided $2,500 to cover costs for which every penny needed to be accounted. ASPP annual meetings typically drew well over 1,000 plant physiologists—an excellent venue in which to promote a developmental perspective. Our three-hour symposium would be held the morning of the first full day of the meeting. This would be an opportunity I would likely not have again. I picked four accomplished speakers who offered a developmental perspective and who represented both top-notch science and topics covering a broad spectrum of research and approaches.

The program committee, a year before the meeting, conveyed their expectations to me for the symposium. "Encourage your speakers to present their talks in a way that can be understood by knowledgeable non-experts and to avoid talking over the heads of the audience. Also, it is helpful if the symposium chair introduces the topic and explains how the pieces (i.e., speakers) fit together." I totally agreed with these suggestions.

I shared these expectations with each speaker and encouraged them to seek out the least expensive airfare and keep other costs to a minimum so that our $2,500 contribution from ASPP would cover most, if not all of our expenses. Each speaker would have 35 minutes to present and address audience questions.

Regarding content, I shared this with the speakers: "Our audience is plant physiologists, most of whom have little grasp of development or genetics. However, they are intelligent, well-trained scientists. We are to provide education about development and illustrate the power of genetics to dissect developmental processes. Use the techniques you know to effectively convey your story: provide background information, define terms, keep data slides simple, use cartoons and illustrations to complement data, and employ summary slides."

Finally, so that I could effectively coordinate the talks, I conveyed this: "Our symposium has a natural flow: meristems make plants so how do they do this [Ed Kledowski, University of Massachusetts-Amherst]? Embryos make the first meristems, so how are embryos made [David Meinke, Oklahoma State University]? Meristems make lateral structures so how are leaves made [Michael Freeling, University of California-Berkeley]? Finally, how are flower parts made [Elliot Meyerowitz, California Institute of Technology]?

"In order for us to develop an integrated symposium, each speaker needs to know what the other speakers propose to say so that each can modify and adjust their talk. I ask that you send to me by June 5 an outline of what you propose to include in your talk. I will then send these (perhaps with suggestions) to all speakers. I will thus know how to introduce the symposium and you how to dovetail your talks."

Professionally and personally, I was deeply gratified by the dynamics of the relations among the presenters. We met for dinner the evening before the symposium that spouses attended. The symposium itself exceeded my expectations. The audience was so attentive that a dropped pin could have been heard. Every talk elicited thoughtful questions and each speaker kept to his allotted time. As a young scientist attending similar symposia, I never expected to be on the stage introducing the esteem presenters.

After the talks, the speakers and I went to lunch and then sought out tourist shops near the Albuquerque convention centre. While browsing an upscale gemstone shop, Freeling, who is a dead ringer for Robin Williams in looks, behaviour, and intelligence, engaged in a deep conversation over several exquisite gemstones with two saleswomen. I overheard one of the women ask the other, "Doesn't he look like Robin Williams?" Freeling was still discussing the magical properties of gemstones when the rest of us headed for the door. "Robin," I said, "time to go." He thanked the women, who were ecstatic about trying to sell a gemstone to Robin Williams.

*****

The second story is an international gathering of Australian and U.S. scientists who researched early events in flowering. Some relevant background about Australia: although it is about the size of the continental U.S., Australia was unknown to Europeans and most of the rest of the world until 400 years ago. Dieppe mapmakers in the mid-1500s placed a giant landmass named Java la Grande between Antarctica and today's Indonesia. It was likely Portuguese sailors who were the first Europeans to see Australia that had been inhabited by humans for tens of thousands of years. However, it was a Dutch navigator, Willem Janszoo, who chanced upon Australia in 1606 and documented the first landing by Europeans. Strange as it seems today, Europeans knew in the 1700s that Australia existed, yet they could not locate it. Sailors did not know precisely where the continent was—nor, in fact, exactly where they were where while at sea. To know one's location, one must know the exact time relative to a given meridian. Global Positioning Systems were centuries away.

Chronometers (a very accurate clock employed at sea) that could function reliably shipboard for long periods in harsh and turbulent conditions did not exist when Europeans first learned of Australia. Because trade and exploration were so important, the British Board of Longitude in 1714 offered 20,000 pounds for a seaworthy chronometer. Two decades later, a British cabinetmaker, John Harrison, submitted his first durable chronometer. Three more followed over the next two decades, all satisfying the conditions for the prize.

Dozens of ships had chanced upon Australia prior to that of Captain James Cook and the *Endeavour* in 1770. By using a seaworthy chronometer, Cook sought and found the once elusive southern continent, making landfall at what was to be Botany Bay and claiming for England all of Australia's east coast. Even today, Australia is a faraway place, somewhat remote and isolated. Had Lloyd Evans, Rod King, and Jim Peacock not invited me, I likely would not have gone, at least not at that time in my life. Maybe never, because I only travelled when I had a professional or important family obligation.

Australia has a population of just over 20 million, most situated near the coasts. As noted earlier, the country has a modestly sized but accomplished scientific community thanks to the historic hiring practices of CSIRO and a handful of quality research universities. For scientific meetings and other professional gatherings, Australian scientists often travel overseas and otherwise seek interactions with foreign scientists. About halfway through my time there, Evans and King posed the idea of organising a joint meeting of Australian and U.S. scientists who were researching flowering. I welcomed the idea. Before I left, we had laid a solid foundation for the gathering.

As for location, Honolulu, located between our two countries, seemed the most obvious. A meeting of 20 to 30 participants would best allow for a breadth of expertise and topics, in-depth interactions, and the effective exchange of ideas. We settled on the title, *Early Steps in the Flowering Process.* Evans was an acknowledged leader in the world of flowering, comparable to Lang, and knew personally the flowering researchers in Australia and other countries. King and I were midcareer and knew many flowering scientists in our respective countries and elsewhere.

King and I began seeking financial support for the conference. He obtained $14,000 (A) from the Department of Industry, Technology and Commerce, and I landed $23,000 (U.S.) from the National Science Foundation. This would cover most of the meeting's costs and scientists' expenses. Even in 1991, these two grants were a modest expenditure for a six-day gathering of 28 scientists. At the end of our nine-page workshop program, I included a comment from Donald Frazier:

No finer use of a grant can be conceived. This puddling of super research brains has a potential of cutting years off of individual research projects, especially if the atmosphere is free of personal or institutional jealousy and undue pride in achievement.

In my experience, most gatherings of scientists are egalitarian. In this regard, it is worth noting that even the most esteemed and accomplished participant from each country—Evans from Australia and Lang from the U.S.—travelled under the same conditions as graduate student and postdoc attendees. Everyone flew economy class and lodged in similar accommodations. Lang did not charge us for the first leg of his flight because he was attending National Academy of Sciences meetings in California prior to the conference. I suspect this frugal behaviour reflects the reality of an academic research scientist's funding: often short-term and with no future guarantee of support. Scientists tend not to waste research funds that are scarce and hard to come by.

I spent tens of hours managing the NSF funds and meeting the requirements of the grant, including publishing the outcomes of the conference, "Early Events in Flowering: A Black Box with a Bright Future," in *The Plant Cell*, a top-of-the-line journal with an international circulation of 5,000. *The Plant Cell* paper was also summarised in the *Flowering Newsletter* published in Europe by the International Working Group on Flowering and in the *Plant Developmental Biology Newsletter* published in the United States. King's summary, "Early Events in the Flowering Process: Key Issues from a Recent Workshop," was published in the *Australian Journal of Plant Physiology*.

Fourteen scientists from each country attended the meeting. Eight were senior scientists; 10 were well established; and 10 were junior scientists, postdocs, or graduate students. Their work represented a dozen plant species reflecting most of the model systems studied then, each focused on a particular aspect in early flowering from a diversity of perspectives and techniques—biophysical, breeding, chemical regulation, developmental, genetic, horticultural, molecular, morphological, physiological—to assess the current state of knowledge and major trends.

Rarely do researchers with such widely different perspectives meet in the same room to learn of each other's work. This meant that each of us had to define unfamiliar terms, state our assumptions, and proceed slowly so that all could understand what was being presented. Most sessions began with a brief overview of the topic and allocated half of the scheduled time for questions and discussion.

The meeting composition and format encouraged reflection and discussion—atypical for the formal parts of most scientific gatherings.

Without question, Early Steps in the Flowering Process was the most dynamic, interactive, and brain-challenging gathering of scientists working on a common process of which I had been a part. My article in *The Plant Cell* concluded with this observation:

What fascinated me were the dynamic interactions among the various disciplines represented at the workshop, which have such profoundly different views and interpretations of the flowering process. Members of one discipline did not appreciate or understand all that members of other disciplines were saying, but they talked and listened to each other with interest and respect. The big problems concerning juvenility, phytochrome-regulated leaf processes, apex induction, organ positioning, and control of gene expression patterns, among others, are still unresolved. However, as was apparent throughout our discussions, the answers will not come from physiological or genetic or molecular analyses in isolation, nor will they come from the study of one organism like pea or *Arabidopsis*. Vehicles at all levels (e.g., undergraduate education, graduate training, professional meetings, funding panels, etc.) need to be enhanced or established that will promote disciplinary interaction, respect, and cooperation. The lack of substantial efforts along these lines will guarantee that many of us will continue to rediscover the known and miss the obvious. (Carl N. McDaniel, 1991, Early Events in Flowering: A Black Box with a Bright Future.)[3]

*****

The only other meetings I have attended that came even close in intensity and quality were Plant Cell and Tissue Culture Gordon Conferences. Gordon Conferences date back to 1931 and have increased in number to more than 200 annually today. Each conference focuses on a relatively narrow area of science that is on the cusp of innovation and major breakthroughs. Gordon Conferences are prestigious international scientific gatherings held at New England prep schools and at private colleges in New Hampshire and Rhode Island. Scientists from academia, industry, and government organise the meetings, which carry a few inviolate rules: participants are admitted by application only, attendance is limited to 125, only 25 participants give formal presentations, presentations and interactions are off the record, press is not permitted, and formal talks are scheduled for morning hours only, leaving five afternoons free for spontaneous interactions. The prep schools are located in small, isolated areas with no restaurants, movie theatres, or other distractions, encouraging graduate students, postdocs, and young scientists to intermingle with the senior scientists whose work they know only from textbooks or literature.

I was accepted to attend the third Tissue Culture Gordon Conference in 1979 and asked to present on positional information in auxiliary bud development. My

---

[3] *The Plant Cell*, 3:431-433.

talk was scheduled for Thursday from 11:30 to noon. Naturally, I was nervous as my time to speak approached; more so because the first speakers used more than their allotted time. I was introduced at 11:50. We had been in the room since 8:30, so I cut my talk short so as not to hold up lunch. I tersely described my node-counting discovery, the bud grafting and rooting results, and the pot-on-pot experiment, concluding with the continual re-rooting experiment showing that when roots are near the bud, it would never flower—according to Lang's definition of never. When I finished, the audience broke out laughing. Confused, I asked, "What's so funny?" which elicited more laughter. At Friday's closing session, people were given various awards. Mine was 'Funniest Person'. I have never before or since been considered funny, much less the funniest!

As you read in Chapter 4, a research university scientist faces two must-do tasks: publish papers and secure funding. When successful, the scientist also takes on the tasks of reviewing manuscripts and grant proposals. I had many mentors, but in particular Anton Lang and Ian Sussex. Both were scientists and persons I came to know well and respect deeply. Yet their approaches and personalities were anything but similar. Sussex, a New Zealander, was open to any idea and far more curious about where that idea might go than judgmental. Lang, born in Russia and raised in Germany, reflected the strictness and preciseness of his heritage. Both men were giants to me. Their influence is woven throughout this book and in my life, professionally and otherwise.

*****

Few people are gifted writers. Many would deem James A. Michener to be one, but like much of what we believe, we would be wrong. In an interview, an enamoured reader asked Michener, "As one of the world's greatest writers, what is your advice to aspiring writers?" Michener responded, "I am not among the world's greatest writers. I am one of the world's best rewriters." Michener's candid response gave hope for me improving my writing.

Scott had hacked to pieces my senior literature review. A postdoc in Spence Berry's lab, while critiquing a chapter of my PhD thesis, became frustrated and returned to me a marked-up draft with the comment, "Your high school writing style will not do."

Folke Skoog, an editor for *Plant Physiology*, after reading a paper I had submitted on amino acid transport, replied, "Can't you just write so an ordinary person can understand what the hell you are saying!" Lang's three-page critiques of my submissions to *Planta* were exactly what I needed. His hardnosed editing and detailed explanations showed me what needed to be done, and my rewriting improved.

Over the years I would think to myself when rewriting, *what would Anton do here?* One of the last papers I submitted to *Planta* had less than a page of comments by Lang.

He closed with, "I don't seem to find much to edit. Have you improved? Maybe I am just getting old." I would like to believe it was the former; Lang had a dead-serious sense of humour that was hard to read at times.

Shortly after the paper Singer and I wrote was published in *PNAS*, Lang and I were together at ASPP annual meeting. I commented, "I am pleased the third reviewer of our *PNAS* paper was finally convinced." Lang responded, "Well, that is one explanation. Could it be that you just wore him down?" Later, in his review letter of a paper that Joan Gebhart and I submitted on floral determination in *N. tabacum* Maryland Mammoth, Lang wrote, "Thanks for the *PNAS* reprint. This now, I feel, reads really short and well! It was a pleasure, and very well worth the time, to help with it."

I have reviewed hundreds of grant proposals and manuscripts and have written many reference letters for jobs and evaluations for promotions. In the later, I would include positive statements supported by specific examples affirming particular statements that differed from letter to letter; however, critical statements were few. Over time, this became almost mandatory. Completely honest evaluations, similar to the type given by Anton Lang, would most likely harm a candidate's case. In fact, when I was up for promotion to associate professor with tenure, Lang sent a letter to the promotion committee. A senior member of the committee told me after I was promoted, "Never have Lang write you a letter again. His letter did not help you." I think that is too bad; Lang's letter provided candid information that enabled people to make the kind of distinctions needed in order to make evidence- and argument-based decisions.

Over the years, my reviews of grant proposals ranked in range from excellent to poor, with most falling in the middle. In my files, I found 120 proposal reviews, most of which I had ranked for the National Science Foundation and the U.S. Department of Agriculture. Most were a page long. I had rated 10 of them as excellent; 39 as very good; 32 as good; 33 as fair; and six as poor. At the time, a successful proposal required very good or excellent ratings from all reviewers. Employing that criterion, I had ranked 40 percent of them high enough for funding. In reality, about 20 percent of NSF and USDA proposals were actually successful.

To better understand quantitatively how I had judged manuscripts, I examined the outcomes for the 48 manuscripts I managed as an associate editor for the journal *Developmental Biology*. I read the plant submissions sent to me by the journal editor and made a tentative assessment before sending to reviewers. I compared my assessment with those of external reviews to arrive at the decision I communicated to the author of record. Of the 48 submissions I managed, four were returned because *Developmental Biology* did not publish descriptive studies. Three were rejected because of poor writing or presentation and 18 were rejected for their low quality of science. Six were accepted after minor revisions, and 17 were accepted after major revisions, which usually required a second review by the initial reviewers. Thus, as an associate editor, I had recommended rejecting 25 and accepting 23—similar to the 50 percent acceptance rate for the journal as a whole.

I routinely added edits and comments directly to the manuscripts that were returned to the author. My reviews were rarely less than a page—often two—but seldom more. Most of the accepted submissions first required major revisions. As an associate editor, my best reviewers functioned like Lang; they offered honest appraisals that assisted in my decision and enabled authors to substantially improve their papers.

The review process behind two of the manuscripts illustrates the value of rigorous peer review. Both papers addressed fate mapping via clonal analysis and were submitted from the same lab. For the first, two reviewers each offered two pages of commentary; the reviewer with clonal analysis experience was far more critical. My assessment letter, which asked the authors to substantially revise their submission, iterated and expanded on the more critical review.

The paper's senior author responded with a three-page letter criticising my assessment. My reply? "Clearly, editors, reviewers, and authors do not always have the same opinions, nor do they always interpret results and factual material similarly. The process of review, in the case of well-done important science, hopefully hammers out acceptable, reasonable agreements and a product which is greatly improved."

The author revised and resubmitted the paper, this time gaining praise from the most critical reviewer: "BRAVO. Much thought and hard work went into reworking this manuscript, and the result is superb… Congratulations on a job well done."

Two reviewers examined the second fate mapping submission. I returned to the author five pages of critique, a two-page letter from me, and the manuscript marked with numerous suggestions. This paper would require major revision. The second extensively changed submission that was accepted included a handwritten note thanking me for my comments. I should point out that the senior author of these papers was a colleague and friend who, early in my career, recognised and promoted the importance of my observation that tobacco plants count nodes. When a scientist reviews a manuscript or a grant proposal, it is not personal. Rather, it is a candid assessment of the science being presented.

*****

Gordon Research Conferences are among the best venues for young scientists to meet both statured researchers and newcomers like themselves. At my first Gordon Conference, I launched an immensely productive relationship with Michael Christianson, whom Peter Carlson had hired as senior research biologist at Zoecon Corporation in Palo Alto. Christianson was exceedingly bright, with a steel-trap memory for all things botanical. Like me, he had a high-energy personality and was equally motivated to improve and broaden the discourse among plant scientists. Our first thought was to push plant development as a topic for a Gordon Conference. In 1984 we began creating a list of plant scientists who demonstrated exceptionally high-quality work. Assisting us was Holly Schauer, manager of NSF's developmental biology grants program and business manager of the Society for Developmental Biology. To gain support, we talked with established researchers Michael Freeling (University of California-Berkeley), Ian Sussex (Yale University), and Joe Varner (Washington University in Saint Louis).

By spring of 1985, Christianson and I, along with Dan Walker from UCLA, had laid enough groundwork to begin seeking input from the larger plant developmental biology community. In September, we sent a letter and questionnaire to U.S. and Canadian researchers asking about the usefulness and logistics of a plant developmental biology annual meeting and a newsletter. We summarised the responses in a follow-up letter in November:

... more than 140 persons responded, and many made detailed and extensive comments. It was very satisfying to have such a huge response... we are on the right track... not everybody identified the same track as the right track, but there were definite trends.

Most respondents favoured the newsletter idea and there were some excellent suggestions as well as cautions... Hopefully we can produce a unique newsletter that you will want to read. In the long run, we can only succeed if you become a contributor as indicated below.

We are a diverse group! The responses indicated that to pick just one annual meeting each year as a gathering place would just not work. An obvious solution is to have a gathering (symposium, workshop, social event, etc.) at as many annual meetings as possible. We believe that the Plant Developmental Biology Newsletter can be a facilitator of gatherings of plant developmental biologists...

... Yes, many respondents posed the question. "Is it possible for people with this degree of diversity to be in the same room together and meaningfully exchange views?" Who knows? But what struck us was the very positive and supportive attitude... It is not clear that what we are going to do will really make a difference. But it is certain that if your positive attitudes are converted into positive actions all of us together can form an exciting communication network.

The leadership of the two major plant societies—the American Society of Plant Physiologists (ASPP) and the Botanical Society of American (BSA)—responded to our request for more inclusion of the plant developmental biology perspective in their meetings and journals. This led to my involvement on the ASPP program committee and the symposium in Albuquerque. It also fostered a BSA Workshop/Symposium, titled *Experimental Approaches to and Analysis of Plant Development*, that I co-organised with Elizabeth Lord, University of California-Riverside. The Society for Developmental Biology was also engaged and sponsored more plant presentations at their annual meeting and several plant-focused pre-meetings.

Over the next few months, Christianson, Walker, and I considered the survey responses and how a newsletter might be structured. We held a two-day brain storming session in February 1986 at Christianson's home in San Francisco. The first issue of the *Plant Developmental Biology Newsletter* (*PDBN*) came out that winter. Christianson, as coordinator, assembled the materials and mailed the newsletter. Walker and I provided input to Christianson.

The issue comprised five sections: News and Views, From the Past, Recent References (annotated), Future Meetings, and Opportunities. The News and Views section of the second issue (Fall 1986) announced a Plant Developmental Biology Pre-meeting at the 1987 meeting of the Society for Developmental Biology. The last page included a questionnaire asking for suggested speakers for two proposed symposia.

The second action called for in the Fall 1986 *PDBN* came from Paul Green of Stanford University, who proposed a way to increase the quality of journal submissions:

One possible function for the *PDB Newsletter* would be to provide arrangements for preprint reviews… editors send me many botanical manuscripts that strike me as never having been read and corrected/improved by anyone outside the laboratory concerned. Thus, I reluctantly have to point out shortcomings of organisation, interpretation, etc. that should have been caught in the factory… It would be nice to have a source of thoughtful critics for such manuscripts… Many of us have a few colleagues to give ideal preprint critiques and perhaps the newsletter recipients can become an extended family to play this role. Each of us could give a key word list, address and phone number, and a master list could be circulated.

This was a superb suggestion that spoke directly to me! I took Green's suggestion seriously and a few years later, in the Spring 1989 issue, gave my testimony:

… I have personally taken Paul's advice. Over the past two years I have had four manuscripts pre-reviewed with the following result: when submitted to journals, all of them have been accepted with only minor revisions. This was not my experience in the past—just ask Anton Lang, Folke Skoog, Paul Green, and others about my earlier submissions.

My procedure: 1) write the best manuscript I can with the help of critical reviews at Rensselaer, 2) pick two off-campus scientists who are knowledgeable in the field (and who are good critical reviewers) and ask them to evaluate the manuscript 'no holds barred', 3) review the manuscript with serious attention to all comments, and 4) have local colleagues look at the manuscript again.

In my experience, the total time to acceptance is no longer than it was before—the time for pre-review is no greater than the time lost to 'resubmission' (and saves the embarrassment of rejection). The final paper, however, has been much better; the pre-reviews have improved the manuscript so that the journal reviewers could really focus on well-presented science and make fine-tuning comments. Two other factors are at play. We had additional time to freely improve the manuscript without having to defend it and, in general, the pre-reviews have been more thorough and candid than journal reviews.

Last, but not least, journal editors and reviewers will love you because you are presumably sending them a higher quality product. It really works. Try it.

The Spring 1990 *PDBN* ran a list of 126 plant developmental biologists, grouped into 11 subcategories ranging from anatomy and ultrastructure to tissue culture, who would provide anyone a pre-review of their paper. By its fifth year, the newsletter comprised 10 pages, not including the six pages listing the names of pre-reviewers. Major categories had become: News and Views, Education (coordinated by Susan Singer from Carleton College and later with Tom Jacobs from University of IL-Urbana), From the Grapevine (coordinated by Ann Hirsch from University of California-Los Angles), Annotated References, Selected References (papers *PDBN* editors considered of wide interest), and The Year Ahead: A Clearing House (a listing of major meetings scheduled).

News and Views ran two stories worth noting. "Fantastic Sums It Up!" reported on the results of an earlier request for donations to finance *PDBN's* publication. From our subscriber list of 317, 120 readers provided $1,889—enough to publish for three years.

Second, under New Agenda, I provided this statement:

As many of you have heard, Marti Couch has decided to change the direction of her scholarly activities. Marti writes in the Indiana University Department of Biology's graduate brochure:

"We are living in a time of profound crisis for the assemblage of biological organisms: mass extinctions, rapid climate change, global alterations of major chemical cycles. As a biologist, to ignore the magnitude of this crisis and continue as if nothing was amiss has become too much of a chore for me. Therefore, I have taken time out from science-as-usual to examine the role of research in my field (plant developmental biology) and the scientific approach in general, in contributing to the way of life leading to this state. It is my opinion that basic science as currently practiced supports the status quo. Since what we have been doing is not working, change is in order. But change how? My scholarship efforts are directed towards analysing the possible alternative ways of doing biological research, and the likely environmental impacts of different scientific disciplines and approaches."

Yes, a substantial change, perhaps even radical. Many of us share Marti's concerns. As we witness the 20th anniversary of Earth Day, we realise things are not going well for the home planet. What should each of us do? Marti had decided upon a direction; read her commentary in the April 1990 issue of *The Plant Cell*. I applaud her actions. She has my support; I believe she can make a difference.

Crouch had earned her PhD at Yale, joining the Sussex lab my second year there. We became great friends who would go a year or two without seeing each other but then end up on a bus together at some scientific meeting. Within minutes, we would be in deep conversations on all matter of subjects. She was one of the first plant molecular biologists to employ genetic engineering in agricultural research and among the leaders in the late 1980s. She was working on canola (*Brassica napus* and *B. rapus*, rapeseed), a plant used commercially for its oil, and consulting for Unilever, whose huge oil palm plantations in the tropics were being used to produce edible palm oil. The company used Crouch's embryo culture research to make oil palm trees genetically uniform so they could be grown industrially on huge plantations. This, however, meant that rainforests were being destroyed and native peoples displaced from their homes. Waste products from processing palm oil were contaminating the water, land, and air.

Crouch had connected the dots to realise that the basic knowledge she helped create was causing destruction of the natural world and the lives of indigenous peoples—exactly the opposite of her intentions as a scientist. She had seen a future she did not want to own. The dominant economically centred science that drove ever more technology could not solve the problems it created. Crouch imagined a different view that was ecologically grounded; she even closed her research laboratory to pursue it. Her innocence gone, she told her story in the pages of *The Plant Cell*, where the previous issue's front cover photo announced the scientific importance of her article within.

Marti's connecting the dots and acting dramatically made her an exceptionally ethical person in general and principled scientist in particular. Few consider thoughtfully the consequences of what they do to earn a livelihood, no less the consequences of how they live. Truth be told we live in a destructive culture and no person or group of persons can exist without being harmful. That notwithstanding scientists have a position from which to do good and bad that mandate forethought and hindsight for what they have done. Nevertheless, taking a holistic perspective, scientists are but one component of our culture that is economically centred. Only when culture becomes ecologically centred will science serve humanity's highest aspirations.

The *PDBN* was egalitarian and published from 1986 to 1994, reaching some 300 plant developmental biologists with a wealth of information on meetings, important new papers, blasts-from-the-past, education at the undergraduate and graduate levels, positions available, discussions of where people publish and why, announcements of coming meetings, and half a dozen surveys on what the newsletter might do and their results. The importance and accomplishments of the *PDBN* in solidifying and enhancing the field of plant developmental biology in the late 1980s and early 1990s were substantial.

At least two suggestions were made in the newsletter that impacted the careers of plant developmental biologists in a positive way. First was the notable improvement in submitted manuscripts from readers whose papers were pre-reviewed, as proposed by Paul Green at Stanford University. Second was a recommendation to reduce from five to three the number of rating categories for grant proposals—fund, fund if possible, don't fund. This was the first of three major recommendations made by the 1988 External Peer Oversight Committee of the NSF's Division of Cellular Biosciences, which I chaired. The new rating system was so much better that I began using it for all the grant proposals I reviewed. I recommended that *PDBN* readers consider doing the same.

Walker helped produce the newsletter for the first two years. With enthusiasm, Christianson and I persisted for six more years. My academic interests then shifted from experimental scientist to the interface between biology and economics, coinciding with changes in Christianson's professional life. After the spring 1994 issue, the *PDBN* slipped quietly into the abyss, leaving the plant developmental biology community enriched.

Chapter 6
Facilitating Learning

Human behavioural patterns honed by evolution enable us to survive and reproduce. Viewed in geological time, humanity—in the blink of an eye—has been uniquely successful in producing a population 100 times larger than that of any other animal species 120 pounds or heavier! We set this record at least 200 years ago when our numbers reached one billion.

Our population number—now almost eight billion—affirms on a global scale the success of our systems of learning. We have not only exhibited phenomenal population growth but also promoted exceptional energy and material increases to support billions of individuals. That is, over the last century educational systems have been quite effective in providing the talent and expertise that enabled the size of the economy to increase some 100-fold.

When born, a human is hardwired to learn. Immediately the new-born suckles, recognises mother's voice, and hears and utters sounds. Within several weeks, infants see and follow objects passed in front of them. Within a year, the child learns to walk, to communicate with words learned from parents and others, and to express a distinctive personality. In no time at all, a learner becomes also a teacher, especially to the parents. A teacher does not cause learning, rather facilitates the process.

My first formal teaching assignment was as an instructor of introductory chemistry and biology at the U.S. Naval Academy. I was enthusiastic and enjoyed the process. Upon returning to Wesleyan for my PhD, I was committed and convinced that I could teach well at the college level. Three more years doing science and meeting Carlson disposed me toward research, while exposure to Sussex's vibrant, spirited group made me yearn to combine teaching and research.

Rensselaer appeared to be my wish granted. Nevertheless, my introduction to a teaching-research university was trial by fire. It did not mirror my experience at the Naval Academy, where I had taught three sections of chemistry or biology each term and had minimal administrative tasks. As an assistant professor at Rensselaer, I taught major parts of two courses each semester along with setting up a research lab and taking on substantial duties within the department. Within two years, the trustees' decision to prioritise research over teaching reached the biology department. Earning tenure and promotions now required research accomplishments, funding from very competitive sources (e.g., National Science Foundation, National Institutes of Health, Department of Agriculture

Competitive Grants Program), and publication in well-regarded journals such as *Developmental Biology*, *Nature*, *Plant Physiology,* and *Planta.* This shift in emphasis likely meant less time and resources I could devote to my courses and my interface with students.

Although teaching became less important in the quest for tenure, I enjoyed crafting and delivering lectures. I augmented my lectures with videos and slides I had taken of animals, cartoons, cells, diagrams, ecosystems, plants, and other things that visually connected students to the topic. In the summer of 1970, I took two courses at the Friday Harbour Laboratory in Puget Sound, Washington—one on marine invertebrates, the other on marine invertebrate embryology—supported by a National Science Foundation fellowship. Photos I took there were among some 500 images I employed during 33 years at Rensselaer. Even my research talks were animated and enhanced by slides of graphs, cartoons, and the beauty of the moths and plants I studied.

Whenever I travelled, I took pictures. Whether it was a professional or personal trip—Australia, Canada, Costa Rica, Czechoslovakia, England, Germany, Grand Canyon, Italy, Paraguay, South Africa, or trips to California, Florida, or across the United States—I put together on return a slide presentation of less than an hour for friends and neighbours, all to share our experiences with others and facilitate learning. Often, we had a second viewing to keep the group size about twenty.

*****

There were three overlapping time periods during which my pedagogy underwent substantial change and I matured as an educator. At the Naval Academy, and for almost two decades at Rensselaer, I mimicked the ways I myself was taught. I taught undergraduate courses in embryology, endocrinology, *Homo sapiens* who are we? Introductory biology, introductory chemistry, and a senior laboratory course. At the graduate level, I taught cell and developmental biology, developmental genetics, mechanisms of hormone action, membrane transport, and plant growth and differentiation. Large undergraduate courses were primarily lectures—sage-on-the-stage—while smaller classes combined lecture and discussion. Graduate courses employed a seminar format where review and primary research papers were discussed.

Without question, student evaluations that were required by Rensselaer and the biology department offer useful feedback in improving the facilitation of learning; however, their accuracy in evaluating course quality and faculty effectiveness is problematic. Student reviews do not measure learning—a serious shortfall to be considered later in this chapter. Evaluations are more likely based on emotions than evidence. This is not all bad. Learning, after all, is intimately related to the interactions between instructor and students. Nevertheless, an emotional response to a situation or event can overshadow everything else. An innocent comment by an instructor could upset students or be misinterpreted. For example, in a large introductory biology class a student was reading a newspaper fully displayed in front of him. I asked the student to put the paper down or leave the class. Several students gave this incident as the reason for their poor instructor evaluation.

I found the most useful feedback came from written answers to open-ended questions I created and employed for my courses: "What did you like best about the course?" "What did you like least?" "How would you improve the course?" The responses enabled me to make adjustments because answers were often relevant.

Early on, students in introductory biology would ask questions for which I did not have answers. That sometimes prompted another query, "Shouldn't you know the answer if you are teaching this course?" I soon realised that I needed to reveal how little I knew about biology.

While reviewing the syllabus, I would casually remark, "I really do not know anything about biology."

Usually, after a short while, a student would ask, "Then how can you teach this course?"

I would respond, "Oh, I meant to say that when compared to all that is known about biology, I know vanishing little. However, I know immensely more than you." After that, no one complained when I did not have an answer. Often, I would encourage the student to seek the answer and share it with the class—few ever did.

I also learned that when my research overlapped with my teaching, the capacity to facilitate learning increased substantially because I acquired depth and nuanced dimensions—historical, human, personal, relational, scale, and temporal—on the topic. For example, when Scott Poethig and I employed clonal analysis to create a fate map of the shoot apical meristem of the germinating corn embryo, the dynamics of shoot development crystallised in my mind. I gained a deeper comprehension of and feeling for the organism. In subtle and boisterous ways, I would insinuate my research topics into the classroom, thereby enhancing the learning I sought to provide students.

*****

Period two in my maturation as an instructor was foreshadowed when Susan Singer, then on the faculty of Carleton College, asked me to spend the spring 1991 term there as a Benedict Distinguished Visiting Professor. I would teach one course and give several campus-wide presentations on topics of my choice. Without laboratory and family responsibilities, I had plenty of time to interact with Carleton faculty and to read. Motivated by a lifelong interest in population, pollution, and more recently climate change, I read or reread *Unsettling of America: Culture and Agriculture* (Wendell Berry, 1977), *Small Is Beautiful: Economics as if People Mattered* (E.F. Schumacher, 1973), *The Death of Nature: Women, Ecology, and the Scientific Revolution* (Carolyn Merchant, 1980), and *The End of Nature* (Bill McKibben, 1989). When I returned to Rensselaer and was asked to serve on the School of Science Environmental Task Force, charged to lay the foundation for an undergraduate degree in environmental science, I was primed for the task. The committee was chaired by Bruce Watson, a quiet, delightful person and a brilliant geochemist in the department of earth and environmental sciences—later a member of the National Academy of Science.

At Carleton, I met Keith Harrison, a professor of English and writer-in residence, who had long been interested in the influence of human habitation on the environment and vice versa. Harrison envisioned the nearby Cannon River as the perfect subject for an interdisciplinary course examining the relations between humans and their environment. However, hard as he tried, he could not gain his colleagues' support. I, however, thought the idea spectacular. On returning to Rensselaer I proposed the same idea—using the Hudson River—to the task force planning the environmental science program. They concurred, and One Mile of the Hudson River became the signature course for the environmental science degree program.

In spring 1992, Doyle Daves, dean of the school of science, asked me to be director of the environmental science bachelor's degree program. "Why me?" I asked. He explained that he wanted a senior-level, hardcore scientist to ensure that the program would be firmly grounded in science, but also someone possessing the energy to launch the new program. Watson was now associate dean of environmental science and urged me to consider seriously the directorship. Both had put a full court press on me.

I knew the basics of environmental science, but not more. Creating the degree program would be a full-time job requiring me to discontinue my flowering research, which was still funded. I weighed the options. My research was at a point necessitating a shift in approach from developmental to molecular, which I had neither the expertise nor strong desire to learn. I also knew that the environmental challenges facing our humanity were far more important.

I accepted the directorship in June. My first order of business was to flesh out recommendations by the task force. The environmental science degree (ENVS) would be housed in the school of science, not within a department, with an advisory board providing input and guidance. Selected for the board were five faculty members from the school of science—one from each department—and two from the school of humanities and social sciences. Over the next three years, we developed and taught four interdisciplinary courses: First Year Environmental Seminar, Introduction to Environmental Studies, Classical Writings in Environmental Science, and One Mile of the Hudson River. I was instrumental in creating each and taught or co-taught all four when first offered.

I knew the ENVS students well. Hours of class discussion and my review of weekly assignments in the Environmental Seminar course gave me insight into each student's strengths and weaknesses, and I tailored my interactions with each student to address them. For at least two reasons, I believe I was a better facilitator of learning with ENVS students than biology students. First, Introduction to Biology typically had 100-plus students. Second, I taught only one class to most biology students.

One Mile of the Hudson River and Classical Writings were also team-taught interdisciplinary courses. The material presented was open to different perspectives and interpretations, many valid. Questions such as "What are the cultural and monetary values of the bald eagle species or of drinkable Hudson River water?" do not have answers as definitive as the molecular composition of a caffeine molecule. Answers were conditional. Effective facilitation of learning required interactive methods more akin to graduate seminars than to sage-on-the-stage lectures.

The story of One Mile of the Hudson River is worth telling. John Schumacher—an exceptional scholar and down to earth human being who held a BS in mathematics and master's degrees in mathematics and philosophy from Rensselaer, and a PhD in philosophy from the University of Oxford—and I co-chaired the Interdisciplinary Environmental Education Committee, which represented relevant disciplines in the schools of engineering, humanities & social science, and science. We obtained a strategic-initiatives grant of $140,000 from Rensselaer to develop and support the course, which included a two-day workshop attended by Keith Harrison from Carleton College who had had the course idea and David Orr who headed up the environmental studies degree at Oberlin College.

The course objectives, as stated in the first syllabus, were in a word, audacious:

We all live in a particular place. Modern life has, to a lesser or a greater degree, removed and insulated most of us from our piece of the 'natural' world. A major goal of the Hudson River course is to acquaint us with our place of habitation. In the process, we will explore how the Hudson River is affected by diverse, interrelated forces and processes acting on it, in it and around it, as well as how the river can be an active participant in the dynamics of change and stasis in our world. We investigate these dynamic forces and processes as they relate to two central concepts: ecosystem function and sustainable human habitation of ecosystems.

Several themes—posed as questions—will unite our quest to understand our place of habitation and forces that shape it. 1) How can the Hudson River be defined? 2) What constitutes a stable, productive Hudson River ecosystem? 3) What constitutes a sustainable pattern of human habitation in the Hudson River Valley? 4) What is the relationship between a sustainable human habitation and a stable productive ecosystem?

As we explore these themes—and address these questions—we will develop critical perspectives on the values and measurements adopted in assessing sustainable ecosystems and sustainable human habitation. We know from the outset that the biophysical, political, economic, sociological, etc. values and measurements can be in conflict. One of the course goals is to be clear about the role of values in determining the fate of the Hudson River and to bring value conflicts onto the table for discussion.

The dozen faculty members who first taught the course had expertise in art, communication, biology, economics, ecology, engineering, history, hydrology, philosophy, political science, and rhetoric. Steve Breyman, a political scientist in the science and technologies studies department, and I oversaw the classes, while the various subject-area experts conducted classes and field trips. No exams were given; rather, students kept a detailed journal of their observations, their collected data, and their analyses of and reflections on the course's activities. The journal entries were expected to reflect and integrate previous course experiences. Breyman and I reviewed and critiqued the journals three times over the semester. Each student then worked on a project, choosing a topic that was local and environmental in focus, which they presented to the class in a symposium at the end of the semester. Final grades were determined by an evaluation of essays, homework, lab reports, and the journal (50%); class discussion and other activities (20%); and the project (30%). The essays, homework, lab reports, and journal entries were ranked satisfactory, unsatisfactory, or not done.

The Hudson River class was radically different from any other course that I had given, or the students had taken. Field trips took up more than half the course. Our first 3-hour class included a visit to the Rensselaer Archives, where students viewed the journals and materials of Amos Eaton, a natural scientist and educator who cofounded Rensselaer Polytechnic Institute in 1824 with Stephen Van Rensselaer. Interestingly, in 1811, Eaton began a five-year jail term, charged with forgery while working as a land agent and surveyor. In jail, he began scientific studies that he later continued at Yale and then Williams College, where Eaton taught and published *Manual of Botany for the Northern States* in 1817. His approach—learning by doing—was novel for its time. He conducted field trips to collect plants and experiments in the lab, much like our course design.

The next two classes were field trips. On a warm Friday afternoon in September, we walked three blocks to the Poestenkill millponds and then in and along the stream to the Hudson River. The Poestenkill is one of the three streams on the eastern side of the Hudson that provided waterpower for early 19th-century mills energising the American industrial revolution in the Troy area. On the west side, the Mohawk River powered Harmony Mills—the world's largest cotton mill complex in 1872 when it began production. Today, rusting pipes and remnant stone structures bear witness to this history, situated in a lush, tree-covered gorge located a mere stone's throw from homes, churches, and businesses that are clustered on the gorge's rim. Our class, on reaching the outlet into the Hudson, ate a picnic dinner while sitting on the concrete rubble of the industrial waterfront. There, we discussed the contrast of nature's recovery in the upper Poestenkill in relation to that of its outlet.

The following Saturday morning, we launched a flotilla of six aluminium rowboats into the Hudson just below the dam at Troy falls—the spot marks the end of the 150 navigable miles above New York City—to row with an outgoing tide the six miles to Albany. Thus, 14 students were introduced to a river and tributary whose history was deeply woven into the community in which they lived.

Over the next 14 weeks, we delved into a range of topics: surface hydrology, ecological food chains, physical and chemical measurements and biota collections, river-floodplain ecosystems, river geology and hydrogeology, the cultures of humans who inhabited the Hudson River valley from hunter-gatherers to native Americans to Europeans, the building and economic importance of the Erie Canal, nature and 19th-century American landscape painting, and the industrialisation of the upper Hudson area and its relation to the environment. The course concluded with a redo of our walk from the Poestenkill millponds to the Hudson River and an essay reflecting upon it in light of our new knowledge.

Students might assume that a college course without exams would be easy, but most considered this class a challenge, in part due to the holistic perspective not offered in other courses. I—who participated in every field trip and class session, kept a journal, and read and evaluated all student essays and journal entries—concurred with this assessment. I coordinated the course two more times, after which the perspectives and topics narrowed. After 20 years, it's still part of the ENVS major, although its structure has become similar to other college courses taught by an instructor.

The demise of the Hudson River course was to sail a pedagogical masterpiece into the doldrums. Keith Harrison's brilliant idea broadened the capacity of several dozen students and a handful of academics to engage with human habitation's myriad perspectives on the place that sustained them. The course was highlighted in newspaper and magazine articles and presented on Public Radio's *The Best of Our Knowledge*. Like so many unique programs of quality the economics foretold what was going to be—One Mile of the Hudson requires too much faculty time and energy.

Three years into the program, we were offering five interdisciplinary courses to 30 environmental-science majors. I had had excellent support from the admissions office for half a dozen years, but that ceased when a new director of admissions determined that too few ENVS majors were paying full tuition. With a rapid decline in the number of majors, the program no longer held promise as a major in the school of science that would annually graduate a dozen uniquely educated, environmentally concerned citizens. In 2000 I resigned as director, ending my second endeavour to facilitate learning. A faculty member in the geology department became director and the program continues with a handful of majors.

*****

My decade with ENVS prepared me for the third and final stage of becoming a more effective facilitator of learning. In 1982, soon after attaining tenure, I proposed that Rensselaer create a biology course for non-biology majors with the goal of all Rensselaer students being required to take biology as they did physics and chemistry. The chair of biology liked the idea. We expected the engineering faculty to be the most refractory group. I met with a half dozen engineers—none responded favourably, but for a reason we had not anticipated.

Their opinions of biology were based on their out-of-date high school experiences decades earlier: "It is not science; it is natural history. Biology is just a collection of unrelated observations. Biology does not have fundamental principles like chemistry and physics. Biology does not have predictive power." We had our work cut out for us to overcome these outdated perceptions of biology!

Changes were underway to our introductory courses, so the biology department was receptive to modifying Introduction to Organisms, the course I had taught since coming to Rensselaer, to Introduction to Biology, a course for non-majors. In fall 1982, I taught the new course to 17 students; the next fall, enrolment increased to 27. Encouraged, we asked the dean of the school of science to advocate for the course and identify it as a highly recommended elective in the undergraduate curriculum. He did, and I began the task of informing science faculty, via this memo, about the new opportunity available to all students.

There is a revolution going on in biology that is going to impact dramatically upon all of us. Its current and short-term technological implications are based upon the establishment of the structure of the genetic material in 1953 and breaking the genetic code in 1961. We can now change the genetic constitution of some organisms at will, and the implications of this ability are mind boggling, perhaps unnerving. But this is only part of the story. The increased use of chemicals as part of our industrial lifestyle (plastics, herbicides, insecticides, food preservatives, chemicals in industrial processes, etc.) has had dramatic effects on all forms of life, including human. Medical technology is rapidly changing. Numerous genetic diseases can be detected prior to birth and in some cases prior to conception. Over 150 new biotechnology companies have been formed in the last five years, and numerous established firms have created biotechnology groups. All our students will be faced with important decisions, professionally and personally, which will have major biological components. It is our job as educators to prepare them so that they can meet successfully the professional and personal challenges of the future.

Biology is no longer a listing of various organisms and their basic characteristics. It is no longer a peripheral discipline in science. Biology will be important in the lives of essentially all our students. It is for these reasons that we should encourage all our students to make biology an integral part of their undergraduate curriculum.

The Biology Department now offers a one-semester course for non-biology majors. This fall, Dean of the School of Science recommended to the Task Force on Undergraduate Education that all students be strongly encouraged to take a course in biology.

Thus, began what became a two decade-long process to incorporate biology in the curriculum of all students. It is of historical interest to note that in the middle of the 19th century, biology was part of Rensselaer's core curriculum in two places. The Department of Physical Geography required Structural and Systematic Botany, Vegetable Physiology, and Geographical Distribution of

Plants in the first year and Distribution of Plants and Animals in the third. In reality, we were simply restoring biology to the core curriculum.

The push for biology in the core curriculum eased off until 1987 when a new dean of engineering appointed a committee to create a three-course sequence to modernise the core-engineering curriculum in chemistry and materials, a sequence that could include a biology course. The idea behind the sequence was to introduce biotechnology, modernise the materials course, and integrate general chemistry into three courses team-taught by faculty from different schools and departments.

I shared with the Chemistry Materials Biology Course Study Committee my belief that biology is fundamental to engineering in ways different from physics and chemistry. The thinking that undergirds undergraduate physics and chemistry coursework is linear. At this level, systems have a small number of variables, and the system's behaviour is usually predictable. Biological thinking, on the other hand, is cyclical, and the number of variables at all levels from molecular to biosphere is unknown. Additionally, biological systems are self-regulating by adapting to internal and external perturbations. The complexity and regulative capacity of biological systems is integrated and controlled via myriad mechanisms. Thus, the thinking required for understanding biology is dramatically different from that for chemistry and physics. However, the complexity, unknown number of variables, capacity to self-regulate, and control mechanisms found in biology *are* traits characteristic of the engineering world.

I further reasoned that by focusing on biological control mechanisms, engineers would be exposed to basic biological concepts in ways relevant to their profession. That is, a solution for a regulatory problem in engineering could be found by reverse engineering a similar biological system. For example, if seeking a feedback mechanism for an engineering system, one could examine the positive and negative feedback for metamorphosis in insects or enzyme synthesis during carbohydrate depletion and loading. Unfortunately, with respect to the curriculum, we again fell short: the biology portion was not included in the revised chemistry and materials courses. Nevertheless, faculty members outside of biology had been exposed to the uniqueness of biological thinking and its potential use for engineers.

Charles Vest, the president of MIT, wrote in his institute report for the 1990–91 academic year:

In a major curriculum development this past year, the faculty voted to establish a subject in modern biology as a new General Institute Requirement for Undergraduates, effective with the class entering in 1993. We believe that we are the first university to recognise, by requiring it of all students, the growing general importance of modern biology and the uniqueness of its methodologies. This will enhance still further the quality and relevance of an MIT education.

So now we had one of the most outstanding engineering schools in the U.S. recognising the value of biology for all its students. Rensselaer was moving in the right direction by beginning to think seriously about requiring biology in its core curriculum. Unfortunately, MIT's action did not inspire Rensselaer to follow through, although Rensselaer's president did send Vest's report to all faculty members.

For two decades, I had been unsuccessful in advocating the importance of basic biological knowledge for all students and was getting discouraged. Then in 1999, Rensselaer hired a new president, Shirley Jackson, who had been chairing the U.S. Nuclear Regulatory Commission. Jackson held a bachelor's degree in physics from MIT and a PhD in nuclear physics; she had been the first African-American woman to earn a doctorate degree from MIT. In no time at all, Jackson called for a life sciences course within the core curriculum. This stimulated conversations across campus, each discipline considering the relevance for their students. If relevant, what should the course content be? This was the real McCoy. I joined with two colleagues in biology and the director of the Centre for Innovation in Undergraduate Education to begin working on materials for what was called Next Generation Studio Biology.

In the early 1990s, studio calculus and physics had replaced the traditional lecture-homework-exam protocol for these subjects. Studio classes were taught in classrooms designed for activity-based, collaborative learning that employed computers and small-group interaction. By using testing techniques such as the Force Concept Inventory, teachers could better measure students' comprehension of topics related to mechanics. Typically, learning is determined by employing inventory scores at the start and end of the course as follows: Gain equals (post instruction score [# correct] minus pre-instruction score) divided by (total possible score minus pre instruction score). Learning gains across the country in traditional physics classes were around 0.15 or 15 percent, whether at a community college or a top-ranked university. At Rensselaer, when physics studio classes with activity-based instruction replaced traditional classes, more than a doubling of learning gain was observed: ~15 to ~40 percent.

In teaching introductory biology, I had always striven to facilitate learning by changing techniques, especially in ways that promoted student engagement. In my first classes, I would give a quiz every Friday. Later, I would assign 10 questions on Friday whose answers had to be turned in personally on Monday—thereby encouraging attendance on Fridays and Mondays. At first, students merely needed to answer each question to earn full credit. To raise the stakes, I asked the course's teaching assistants to grade one question from each paper; only a correct answer would earn full credit.

When I first started teaching, my exams were a mix of short answer essays; a long essay; and fill-in-the-blank, multiple choice, and definition questions. After a few years, I determined that a better measure of student understanding would be gained through essay questions only. But in time, the weak writing and dearth of facts disappointed me. The students felt there was too much material to learn for each exam, so I provided beforehand six essay questions that focused on important concepts. Two would appear on the exam. The students' writing improved, as they now had time to research and prepare their answers. I do not know if any of these procedures improved student learning. However, as mentioned above, physics instructors in the 1980s began to use well-designed concept inventories to provide a scientific way to determine if a change in instructional method improved learning.

The new life sciences course at Rensselaer provided a unique opportunity to compare learning gains in an instructor-centred, lecture-homework-exam format with those in a student-centred, Web-based interactive pedagogy. I would be teaching Introductory Biology in the fall of 2003 and Next Generation Studio Biology in the fall of 2004. Both would cover the same material and concepts in evolution and ecology. In late 2002, the Live Sciences Course Committee began creating concept inventories even before the course content had been firmly established, because we were certain that evolution and ecology would be included.

In April 2003, the dean of undergraduate education appointed an advisory committee for the Life Sciences Core Course. Representatives from the Schools of Architecture, Engineering, Humanities & Social Sciences, Management, and Science would oversee the Life Sciences Course Committee, which had been at work for months. By late spring, after countless meetings, conversations, memoranda, emails, and presentations, the committees agreed on the foundational areas that would be of the greatest importance for basic competence in the life sciences: evolution, genetics, molecular and cellular biology, and ecology.

Evolution as the uniting idea in biology was succinctly described by Theodosius Dobzhansky in the title of his famous 1973 essay published in *American Biology Teacher*: *Nothing in Biology Makes Sense Except in the Light of Evolution.* In fact, it is not an overstatement to replace 'in Biology' with 'in the World'. This broadening of the concept of evolution beyond its narrow biological definition—a change in gene (allele) frequency in a given population—to include human culture, technology, religion, and beliefs, reinforces 'change' as a universal aspect of life as we know it. Evolution finds its expression in the biological arenas of cellular, developmental, ecological, genetic, and molecular biology. These processes and the principles that undergird them are fundamental to humanity's survival because they enable an understanding of what we are, where we came from, and our relationship with the rest of life. Humans are just one species evolved over a vast period of time and among millions that make up the ever-changing fabric of biodiversity on which humanity, and all forms of life, utterly depend.

The Life Sciences Committee believed that the new life sciences course would be taught in a way that reflected the timeless pedagogical quote of Confucius: "Tell me and I will forget; Show me and I may remember; Involve me and I will understand." People learn best by discovery and remember what they have learned through repetition and iteration in different venues. The new Web-enhanced interactive Biology 1010 (WEI-1010) built these realities into a structured program of pre-, in-, and post-class materials that emphasised concepts and interaction and encouraged students to discover biological concepts and make them their own. We employed eight venues for interactive learning. This student-centred, Web-based interactive pedagogy differed radically from the way I was taught and how I had taught intro biology courses for 30 years (T-1010). The relevant question was: Would this method improve student learning?

By the end of summer 2003, the Life Sciences Course Committee had identified 47 core evolution concepts and used them to create a 20-question evolution concept inventory; 38 core ecology concepts were used to make a 17-question ecology concept inventory. These inventories were used in the traditional course (T-1010) in fall 2003 and in WEI-1010 in fall 2004. I first taught the pilot for WEI-1010 in spring 2004, so I was familiar with the new protocol. Students in T-1010 and WEI-1010 had a similar knowledge of biology. Both classes were taught the same evolution and ecology concepts, yet the material was presented in very different ways. A significant increase in learning gain was observed with WEI-1010 when the concept inventories were given in the last week of class and unannounced (evolution: T, 0.10; WEI, 0.19; $p = 0.024$ and ecology: T, -0.05; WEI, 0.14, $p = 0.000009$; where p is a statistical indication of significance, the smaller the p the more significant). These results affirm, as in physics, the case for enhancing or replacing instructor-centred lecturing with interactive, student-centred instruction.

We also made the assessment in WEI-1010 using the final exam. We observed significant increases over learning gains measured in the last week of class that occurred without studying or credit: evolution (up 95 percent from 0.19 to 0.37, $p = 0.0001$) and ecology (up 143 percent from 0.14 to 0.34, $p = 0.000003$). The significant increases affirmed that for-credit work does inspire studying and motivation to do well, as most people have assumed for a long time.

Many educators before and since our demonstration have established that interactive, student-centred pedagogies do, in fact, facilitate learning better than sage-on-the-stage. Why, then, does lecturing remain the mainstay in most high schools, colleges, and universities? Of course, there are many reasons. First, the inertia for doing as-has-always-been-done is enormous—big rocks at rest stay at rest. Second, creating quality interactive, student-centred content takes time, resources, long-term commitment upfront and requires maintenance and improvement. Third, education in the United States is beholden to a political market economy that drives all things under the sun to be valued in monetary terms—and lecturing is more economical. That said, effective learning and a durable culture are unlikely to be grounded in economic terms. Important cultural values such as trust, cooperation, fidelity, respect, honesty, and even informed electorates are not explicitly valued in our political market economy.

*****

This brings me to the final topic in this chapter: cheating. Integrity is the *sine qua non* of science culture and the academic world. From as early as I can remember, my parents and the culture in which I matured have valued honesty—truthfulness always; never lie in any important context. I have always been honest and forthright. Mary's older brother once said to her, "I cannot believe you are dating such a straight arrow."

This character trait was solidified, much like my aversion to smoking and drinking, by an event in high school. As a sophomore taking a Latin exam, I had just one question left when the bell rang—a fill-in-the-blank asking for a verb tense. The teacher asked us to pass our exams forward. At that moment, I happened to see the answer on the exam of the student next to me. I wrote her answer on my exam and passed it forward. As I walked out of the classroom, I felt terrible. I knew I had done the wrong thing. I did not have the courage to confess to the teacher, but I was miserable for the rest of the day. I vowed then and there not to cheat in non-trivial matters. Yes, many social situations require grace and compromise; however, to function well societies require fundamental trust and honesty.

Adherence to academic honesty was expected at Oberlin College when I attended. The college had an honour code administered by students—every student was charged with adherence to the code and its enforcement. Exams had to be given without proctors, and on all exams or assignments a student wrote, "I have neither given nor received aid." followed by the student's signature. Today the statement is "I have adhered to the Honour Code in this assignment." The Honour Code is still student run; if a student witnesses cheating, the observing student is responsible for telling the violator to report the incident to the student honour code committee for adjudication. In my four years at Oberlin, I never witnessed nor was aware of a violation and recall the number of incidents annually reported was small, probably half a dozen.

Rensselaer did not have an honour code. My introduction to student cheating was trial by fire. While administering an exam in fall 1980, a proctoring teaching assistant alerted me that two students in the back of the room were behaving as though they were exchanging answers. I watched the students. They appeared to be interacting, but I was too far away to hear or see anything definitively. When the exams were collected, I marked the two so I could identify them after grading. On each exam, the same seven questions were answered wrong, and with the same incorrect answers. The pages also displayed large crossed-out letters in the margins and on the backs.

The probability of having seven wrong—and identical—answers is very close to zero. This appeared to be a case of collaboration, but I needed to speak with the students. The teaching assistant of their lab section told the students to see me the next day. The students did not wait; they came to my office right after their lab. I confronted them with the evidence, but each categorically denied collaboration. I spent an hour carefully reviewing with them the improbability of answering the same seven questions incorrectly and identically. I even told them that none of the other 115 exams had the same combination of right and wrong answers for two of the questions—I had checked. The students stonewalled me; no way would they cheat when both were in the process of interviewing for medical school. Because this was my first encounter with possible cheating, I sought advice from the dean of students, who oversaw instances of academic dishonesty.

Midmorning the next day, one of the students came to see me. Clearly distressed, he admitted that he and his colleague had cheated on the exam and lied the night before, but that he had learned and would never cheat again. He pleaded with me to not report the incident to the Judicial Board, but to give them a zero grade on the exam.

My reply to him: if I had believed their lies, would he be here now admitting he had lied, and that dishonesty bothered him? He said, "no." I told him I would think about the course of action I would take and asked him to come back in three days. In earlier consultations with the dean, we had agreed that the students needed time to think about what they had done. In several days I told them that I had reported them.

Three weeks later, the Academic Judicial Committee for the School of Science met to hear the case. The committee was composed of the associate dean of the School of Science, the dean of students, a physics professor, a student appointed by the Judicial Board, and a student appointed by the Grand Marshall. It was incredible! Although the evidence was solid and the students had confessed to lying, the procedure focused on my actions, as if I were the one who had cheated and lied. What was going on? Had I violated some unspoken rule of not bringing cheating and lying into the light? As we walked out of the hearing, the dean of students quietly remarked, "You would think you were the guilty one."

After this first instance of dishonesty, I brought forth 27 more cases while at Rensselaer. They ran the gamut: plagiarism on a book essay (3), on an exam (2), and on homework assignment (16); use of a crib sheet during an exam (2); use of unreleased library copy of homework answers (2); a request for regrading an exam after changing the answers on returned exam (1); and falsifying results for laboratory experiments (1). I never sought out cheating, but when the possibility was pointed out to me, usually by a teaching assistant, I examined the evidence carefully until I was absolutely certain that a hand had been in the cookie jar. If uncertain, I would talk with the student(s), explaining that I did not tolerate dishonesty nor should they, because without honesty, trust is betrayed, and human relations fall apart.

Rensselaer administers a confidential survey to seniors before they graduate. The answers are collated and available internally. I remember being told that some 75 percent of students admitted to cheating once; around 50 percent had done so twice. Several times I asked the responsible dean how many academic dishonesty cases were annually brought to his office—about six, were his answers. Although hard to believe, this number may be high. A memo dated November 8, 1982, to faculty members from the dean responsible for academic judicial hearings listed all of the hearings from 1978 to 1982—my first case was there. The total was 12, or 2.4 annually. As stated above, during my 33 years at Rensselaer, I brought forward almost one case per year. At face value, this meant I had about 40 percent of all academic dishonesty cases, which had to be wrong. Apparently, most faculty members handled academic dishonesty cases on their own, likely not believing it worth their time to pursue cheating incidents according to the rules. This was unfortunate; it meant the institute was neither treating cheating uniformly nor acknowledging in a public way the extent of academic dishonesty.

In my first case of cheating, I was treated as if I were the guilty party. I made the decision to retire and move to Oberlin in search of a new life when a case of staff plagiarism was handled in a similar way. An examination of how institutions of higher learning handle lapses of academic honesty by faculty and staff forthrightly belies that the *sine qua non* in academia is integrity. The history of whistleblowing throughout the U.S. and other cultures illustrates the overwhelming inertia that leads us to shun whistle-blowers. It can be argued that this is a deeply ingrained element of human nature, because to go against one's group is to threaten the group itself. Knowing human behaviour was honed to foster survival of the group when we were hunter-gatherers indicates why shunning whistle-blowers is to be expected. I knew that to formally expose an instance of staff plagiarism was risky business. Nevertheless, to do otherwise was not possible for me.

The process was rigged from the beginning when members of the committee appointed to evaluate my charge had conflicts of interest with the accused. The committee dismissed the plagiarism charge. When I appealed the decision, because of conflicts of interest, I was told that the decision could not be appealed. It was time to retire and create a new life.

Chapter 7
Community Service

John Gowdy walked into my laboratory announcing, "Here it is! Our Easter Island!" while handing me a *New York Times* article titled, *A Pacific Island Nation is Stripped of Everything.* This was December 1995, and I was in the midst of winding down my developmental biology research program and bringing up to speed Rensselaer's new environmental science program.

Gowdy and I had gotten to know each other two years earlier at a seminar called Environmental Issues: Economics, Technology, and Policy, held by the Capital District's higher education institutions, all of which were launching environmental programs. As the representatives from Rensselaer, Gowdy and I decided to carpool to Siena College for the weekly sessions. We hit it off instantly! He read *Nature* and *Science* for the environmental articles, while I strove to understand the total mismatch of the natural sciences with the core principal of economics—unlimited growth: a concept not tenable on a finite planet.

After a dozen drives together, we resolved to write a paper on the most important challenge facing humanity—our ever-growing human population and associated excessive consumption of resources: fish, forests, minerals, soil fertility, and occupying space thereby denying habitat for other species. We soon recognised, however, that population itself was not the challenge; more problematic was our diminishing biodiversity and the life support it provides. Clearly, if life support were unaffected by 50 billion people, then Earth would be able to support many more than our current 7.5 billion. But it cannot! Humans use 40 percent of net terrestrial productivity. We occupy or otherwise use a huge fraction of the space that previously harboured our most productive ecosystems. We also cause worldwide pollution, alter global chemical cycles, and significantly increase atmospheric heat-trapping gases—all drivers of the sixth mass extinction of life that is currently underway.

In 1995 Gowdy and I published *One world, one experiment: addressing the biodiversity-economics conflict* in *Ecological Economics*. We spent a full year writing the article with a dozen colleagues pre-reviewing it. Fundamental conclusions of the paper were succinctly stated in the abstract:

The self-organising principles of markets that have emerged in human cultures over the past 10,000 years are inherently in conflict with the self-organising principles of ecosystems that have evolved over the past 3.5 billion years. The rules governing the dynamics of ecosystems, within which all human activity takes place, are ultimately a function of biological laws, not the function of human-created economic systems. The conflict between these systems is illustrated by the fact that economic indicators have shown vigorous growth during the last century while a variety of environmental indicators have exhibited negative trends. Ultimately, however, the growth of human economies faces the constraints that limit all biological systems. In this article, we outline the bases for the conflict between biological and economic activity and suggest policy approaches that will enhance the chances for creating cultures that are economically and environmentally sustainable.

Simply put, biological laws will prevail, and change will be forced upon human-devised economic systems. The sooner that humanity accepts this ecological perspective, the sooner our societies can attempt to create durable patterns of living.

Gowdy is a prolific writer with a sagacious interdisciplinary mind. Our relationship would shift my academic interest from experimental science to the interface between biology and economics. We wrote two more articles for economic journals, which led to the high point of our collaboration—a book, *Paradise for Sale: A Parable of Nature.* This was the subject of the *New York Times* article that Gowdy had handed to me. The Pacific island that we wrote of—a nation 'stripped of everything'—is Nauru. The story of Easter Island, whose culture came undone a century and half ago, is known by many people and should have been an unremitting warning about the collapse of a civilisation catalysed by its excessive exploitation of natural resources. But it is not! People today dismiss the undoing of Easter Island because its inhabitants lacked the knowledge and technologies available today. Nauru is the story of modern civilisation writ small. It cannot be dismissed easily.

Located between Australia and Hawaii, Nauru lies 15 miles south of the equator in an isolated expanse of ocean, thousands of miles from anywhere except Banaba, a small island of similar origin located some 300 miles east. Because of strong westerly currents and winds, a few people from Banaba were able to reach Nauru by chance but traveling by canoe in the opposite direction is not possible. Both are seamount coral islands that possessed immense deposits of highly enriched phosphate. Although each island has a story worth telling, Nauru's is the most relevant today.

Some 3,000 years ago, Micronesians, most likely lost in the western Pacific, chanced upon the island. Their settlement of it created a unique language and culture. A whaling ship, *Hunter*, came upon Nauru on November 8, 1798. Sailors did not leave the ship, nor did Nauruans board, but the ship's captain gave the island its English name—Pleasant Island—based on its appearance and the islanders who paddled their canoes out to see the whaler. The first beachcombers took up residence in 1830, thus beginning the influence of western culture on Nauruans. Soon Europeans controlled the island. Phosphate—an absolute necessity for agriculture—was discovered on this island at the turn of the 19th century and mined by the British-run Pacific Phosphate Company.

With the end of World War I, Nauru, which had been under German control, was put in a League of Nations trusteeship administered by Australia, New Zealand, and the United Kingdom until Japanese forces occupied the island during World War II. Nauruan culture at the time was vibrant and mostly intact, but the Japanese occupation radically impacted Nauru and its people. About 1,200 Nauruans—two thirds of the population—were taken to Truk atoll, where some 400 died under forced labour. Those who remained on Nauru also fared poorly. Allied forces cut supply lines and routinely bombed the island. About 5,000 people, mostly Japanese, were trapped on an island with a carrying capacity of about 1,000. Everything edible was ravaged; starving people were surviving only on pumpkins grown in 55-gallon drums, their vines fertilised with human faeces and urine. When Nauru was liberated and those on Truk returned, the island and culture were beyond recognition. Nauru was put under a United Nations trusteeship, again administered by Australia, New Zealand, and the United Kingdom.

The British Phosphate Commissioners, who had replaced the Pacific Phosphate Company after WWI, revived the phosphate industry that provided work for most Nauruans. 'Pleasant Island's' forests, coconut groves, homes, infrastructure, and economies—traditional and Western—were in tatters. In 1933, fewer than 10 percent of adult males held jobs; in 1948, almost 90 percent did because the British Phosphate Commissioners provided the only livelihood.

The small amounts of money that some Nauruans acquired from phosphate mining prior to WWII had introduced them to the market economy. Some acquired tastes for tobacco and western foods such as biscuits, canned salmon, flour, rice, and sugar. After the occupation, all Nauruans were fully engaged in the global market economy.

The United Nations in 1947 had not intended a permanent trusteeship for Nauru, despite considerable efforts of the British Phosphate Commissioners to the contrary. On January 31, 1968, Nauru gained its independence and control of the remaining two-thirds of the island's phosphate that had not yet been mined. After 67 years of control by others, the Nauruans were again fully in charge. However, they were a mixture of who they used to be and what they had become. Nevertheless, Westerners had instructed them, directly and indirectly, on the workings of the global market economy. Nauru thus became a test tube experiment on the durability of the current western form of civilisation. The hypothesis to be tested: is our current capitalistic market economy durable?

Soon after liberation, the Nauruans were enjoying the material things that wealth provides—cars, education, radios, refrigerators, televisions, and travel. In little more than a decade, the Nauruans were, per capita, the wealthiest people in the world. Nevertheless, Nauruan leaders recognised that their supply of phosphate ore was finite and by the mid-1980s had placed over a billion dollars (about $200,000 per Nauruan) in trust funds that made investments around the world. These included Nauru Pacific Shipping Company; Air Nauru; Nauru House in Melbourne; buildings and hotels in Guam, Honolulu, Houston, and

Washington D.C.; and in such varied enterprises as forest land in Oregon, a brewery in the Solomon Islands, and a phosphate plant in India.

Managing such diverse, worldwide investments took knowledge, skill, wisdom, and luck. Scam artists abounded, and markets experienced highs and lows that were difficult to navigate. By the late 1990s, when Gowdy and I visited Nauru, the easily accessible phosphate was nearly expended, and the vagaries of world markets had chewed up most of Nauru's nest egg. Broke soon after the turn of the 21st century, Nauru resorted to all manner of schemes to generate income: charging fees for housing refugees for Australia, for Nauru passports, and for fishing rights. Some individuals established electronic banks without records to launder enormous sums of money for considerable profit.

Nauru's biodiversity, too, was devastated. No records exist of the island's biodiversity prior to its western discovery in 1798, so it's not possible to assess the consequences of human habitation over the first few thousand years. However, based on similar situations, we presume that some species went extinct as ecosystems were influenced. That led Nauru's stable culture and population of around 1,000 to come into equilibrium with the island's biota. The invasion of Europeans and alien species extensively changed the biota, with the flora being indicative. Most of the native flora was lost—55 of 487 plant species on Nauru are native, with 28 of the 55 being rare or likely extinct. More than 50 percent of the non-natives are ornamentals, and some 20 percent are weedy species—meaning that most of the native habitat is gone. The scientists who generated these data concluded: "Given no deliberate human intervention, the succession to the potential natural vegetation and an environment capable of sustaining human life will take many thousands of years."

The story of Nauru provides a test-tube experiment on the viability of our current capitalistic economic system. The Nauruans disproved the hypothesis that the present economic construct is durable. Even before this elegant test of the hypothesis Gowdy, many others, and I had plenty of evidence, provided at the conclusion of this chapter, that our current economic system is doomed to collapse civilisation as we know it.

*****

Population was an issue that had long attracted my attention. In the early 1980s I became aware of the Population Institute (PI), a non-profit whose mission was to raise awareness and marshal U.S. resources to meet requests from third-world nations seeking assistance with population management by providing the knowledge and wherewithal for women to regulate their fertility. Werner Fornos ran the organisation, which depended heavily on student interns to plan, design, and implement programs. Among them was the Educate America Campaign, in which Fornos visited seven states a year for weeklong events. I learned of his visit to New York in the mid-1980s and invited him to speak at Rensselaer. I reserved a classroom with 225 seats and promoted the event at Rensselaer and the nearby Russell Sage College with fliers, personal notes to faculty, and college event calendars. Fornos, who had recently broken his ankle, limped in on crutches to an audience of just 22. Still, he delivered an animated, energy-charged presentation and answered questions. This was my first time inviting a statured, internationally known person to speak at a campus-wide event. I was deeply embarrassed by the sparse turnout. I expressed as much to Fornos as he was leaving. Without mention of it, he smiled and thanked me, commenting on his next event—a local TV news interview—as he hurried off as fast as his crutches allowed.

Seven years later, Fornos returned to New York and to Rensselaer for a second presentation. By this time, I had learned how to guarantee an audience. Per usual, I was teaching an introductory biology class with well over 100 students. Announcing the Fornos talk, I encouraged students to attend, adding that a one-page commentary on the talk would earn them 10 extra credit points. I encouraged my colleagues to do the same. The 200-seat room and side aisles were packed. Fornos again presented a high-energy, informative talk, followed by 20 minutes of questions and a line of a dozen students waiting afterward to speak with him individually. As we left the room, I thanked him. With a smile, he responded, "You have recovered nicely."

In August 2002, Fornos invited me to serve as a volunteer in the Population Institute (P.I.) delegation to the second United Nations World Summit on Sustainable Development (WSSD) in Johannesburg, South Africa. I hesitated, as the late summer dates coincided with the first two weeks of classes. The first weeks of a large introductory course were most critical, and I had no qualified substitute for my Introduction to Biology course. Fornos asked me to think it over and get back to him the next day.

I told Mary about the invitation that evening. "Go!" she said. "What a great opportunity." We took a walk to talk it over—that's what we did when weighing important decisions. On the way home, we stopped to visit a good friend and Oberlin graduate. She immediately responded, "Of course you're going. Rensselaer would be proud. You can figure out how to handle your class." As Mary and I walked the several blocks home, I hit upon a solution: I would videotape the six lectures and ask a colleague to show the videos and answer student questions. He had been an older student in the course a year earlier and agreed to help. The situation was not perfect, but it was workable.

The PI delegation consisted of four staff members and nine volunteers—many on the older side. Eight of us were over 60, two under 30, and three in between. We flew together on the 18-hour flight from Atlanta on August 25, arriving in Johannesburg the next morning. Some 22,000 people participated, including 104 heads of state and country delegations. The goal of this WSSD was to consider what had been accomplished since the 1992 Earth Summit in Rio de Janeiro and to build upon it. Secretary of State Colin Powell was the only U.S. government official to attend; he stopped briefly to speak at the end of the summit. And President G.W. Bush? He was spending time at his Texas ranch cutting brush. PI was the only United Nations-registered non-governmental organisation (NGO) to represent the issue of population-consumption relative to sustainability. As a super minority, Fornos' delegation assumed an oversize role.

Our charge was two-fold. First, we interviewed delegates to provide stories for PI's *Hometown News Service* program (*HNS*), which had been part of all major international conferences focused on population and sustainability issues since the 1974 World Population Conference in Bucharest, Romania. Second, we secured signatures on a petition to put overpopulation on WSSD's agenda.

HNS can be summed up in two words: elegant simplicity. We asked delegates to fill out a form asking basic information—name, address, rank/title/position, and hometown newspaper. The form posed two questions: 1) What impact will this conference have on the future of the planet? 2) What role do you think rapid population growth plays in sustainable development problems? In exchange for the information, delegates received an attractive canvas bag with 'The Population Institute' under the WSSD logo along with information about The Population Institute on one side all in blue. On the other side was 'A World of 6.2 billion and growing…' in blue above 'HOW DENSE CAN WE GET?' that is red. Using the answers to the two questions, PI staff then drafted news releases for the delegate's hometown newspaper that highlighted the delegate and the population-sustainable development issue. By Fornos' estimate, the program had provided over the years thousands of articles to local newspapers across the planet.

Our second task was to obtain signatures on a petition addressed to the leaders of WSSD related to the issue of overpopulation. The document outlined the current situation—that world population had doubled since 1960, global consumption had increased twofold since 1970, 60 countries were unable to feed themselves, and 100 million people lacked safe, fresh water. The petition asked the summit to endorse the 1994 Cairo population and environment agenda and the Rio 1992 demographic statements. In this way, overpopulation would be part of critical conversations addressing issues such as inequity, poverty, education, clean water, and disease.

While the volunteer delegates handled these tasks, Fornos and his staff were politicking influential people in the UN and government to give PI's message a presence at the summit.

Fornos had arranged for us to stay at Palm Grove Lodge, about 15 miles from Sandton, where the summit was happening. We realised quickly that our trips to and from Sandton, which required taxis and bus rides and took two hours each way, would lessen our time at the summit. We needed to improve our options. A fellow volunteer who had arrived in Johannesburg a week earlier found a taxi service, which Fornos hired for one round trip per day for each PI delegate. A huge improvement!

I headed off the morning of the second day with 25 bags, 25 invitations to a reception that PI would be hosting at the lodge Thursday evening, a stack of *HNS* interview forms, and a *Christian Science Monitor* op-ed that Fornos had written: 'The People Equation' led off with a quote from eminent Harvard biologist Edward O. Wilson: "We are inside a bottleneck of overpopulation and wasteful consumption." Fornos concluded the piece with his own observations regarding the preparations for WSSD: "There has been a strange silence on population issues that are at the very core of the coming deliberations."

I had no problem handing out the invitation or the op-ed; however, I did not like walking up to people and asking them to fill out the *HNS* form—it made me feel like a door-to-door salesman. It was a slow process. I obtained 19 *HNS* forms the first day. Lots of people turned me down, including Christian fundamentalists and others who did not view population as a problem. The next day was no better—18 forms. On Thursday, several of us set up shop in the NGO information room that people passed through on their way to conference rooms. Many people were attracted to the canvas bags; others came to see what the crowd was about. I easily acquired 32 *HNS* forms that day and the next, with the numbers climbing to more than three-dozen on each following day.

We travelled to Cape Town for the weekend to holiday and purchase souvenirs. Fornos had learned long ago that "all work and no play" was not effective for recruiting volunteers who paid their own way. He arranged a tour to Seal Island, the Cape of Good Hope, and Cape Town's botanical garden. The Cape Floral Region situated in southern South Africa is exceptional and spectacular. Twenty percent of Africa's flora grows there, making it among the world's most botanically diverse areas. Almost 10,000 species can be found in 30,000 square miles, making it the smallest of the world's six recognised floristic provinces. About 70 percent of the species are endemic, meaning they are found nowhere else. Fynbos, perhaps a useful scrabble word, is the name given to the Cape Floral Region biome. The tour also took us to a beach of jackass penguins that came to the beach some 18 years ago. Named for the jackass-like sounds they make.

Back at WSSD on Monday, we moved to a new location, where I found a table leaning against a tent. I rolled it into an open space and set out our materials. We soon had a curious crowd, including South Africans, many of whom were workers at the summit, who wanted the canvas bags. They volunteered to fill out the *HNS* sheet and tried to convince us they were not from South Africa. We explained our purpose—that we were raising awareness of population and sustainability and that we had already interviewed many South Africans. I felt terrible; these people could have used a quality bag and by not interviewing them, we had created bad feelings. I was the 'ugly American!' Two other PI volunteers had also been overrun by South Africans in a get-a-bag stampede. Other than this friction, the South Africans were superb hosts.

Tuesday, we all dressed up—sports jackets, slacks, polished shoes—to hand out a PI press release that highlighted the population-sustainability connection and its contrast with the actions of wealthy countries, which "coddle the comfortable and ignore the afflicted," as Fornos wrote. Members of our delegation stood at the entrances to the main conference centre and distributed 2,000 copies of the press release. It was effective in securing a full column in the *Summit News* and mentions in other media. When we finished distributing the op-ed, I walked to *Time* magazine's location in the conference centre to get 150 copies of its World Summit Special Report, which I had read on the flight to Cape Town. It was quite good, and I wanted to give a copy to each student in my introductory biology class and to an Oberlin alumni affiliate group that I chaired.

Our mission at WSSD ended on Wednesday, September 5. We had generated 660 hometown news releases and accumulated several thousand signatures on the petition, which Fornos delivered to the conference director general's chief of staff. The world leaders attending WSSD affirmed the linkages between poverty eradication and environmental protection; health initiatives, including family planning and reproductive health; and equity, equality, and the empowerment of women. Perhaps the petition had helped add population connections to the goals of the summit.

Concern about our ever-growing population is a topic with a long history. In 1798 Thomas Malthus published the first notable global call for concern in *Essay on the Principle of Population.* Modern equivalents came in 1968 with Paul Ehrlich's *The Population Bomb,* followed in 1990 by *The Population Explosion,* written by Paul and Anne Ehrlich. These writings raised significant public awareness of the consequences wrought by billions of people.

Knowing this history, I would give good odds that Fornos has been among the most influential advocates for addressing the world population challenge in the past 100 years, if not the most influential. Consider his tireless three decades at PI, where he produced programs such as *Hometown News Service*; the Educate America Program, in which he visits seven states each year, each for seven days to hold seven events per day; the We Care network of volunteer state and congressional district coordinators; World Population Awareness Week; *POPLINE*, a bimonthly publication dealing exclusively with international population and development that is sent to 2,000 newspapers worldwide; and Global Media Awards. Fornos is also a persistent lobbyist in Washington, D.C., and hundreds of interns have passed through his organisation.

*****

Unfortunately, the population-consumption issue rarely is given the funding and media attention of other global issues. As the year 2000 approached, for example, there was growing worldwide concern that computer systems would malfunction with the year change. The *Washington Post,* in a November 18, 1999, article titled, *Y2K Repair Bill: $100 Billion,* stated:

U.S businesses and government agencies are being forced to spend about $100 billion to keep the year 2000 glitch from crashing their computers, making a simple two-digit programming 'bug' the most expensive peacetime catastrophe in modern history. [4]

It turned out to be a non-problem. Computers behaved normally on January 1, 2000, and throughout the year. Most of the $100 billion was spent ahead of time to 'fix' computers to recognise a four-digit year date. Yes, if computers

---

[4] www.washingtonpost.com/wp-srv/WPcap/1999-11/18/077r-111899-idx.html

everywhere had gone haywire, chaos would have reigned. Nevertheless, why did the United States spend comparatively a pittance for Y6B, the year of six billion people? For more than a decade, demographers had predicted that the human population would race past six billion in October 1999—just 12 years after reaching the five billion mark. This would lead to innumerable catastrophes, many of which have arisen; famines, water shortages, overcrowding, local conflicts, refugees, and accelerated climate warming. As many experts have argued, almost all challenges faced by humanity in the 21st century would be easier to meet with fewer people.

The Population Reference Bureau and the National Peace Corps Association created an event titled *The World of Child Six Billion.* This inspired me to organise 'Rensselaer County in a World of Six Billion People' for the week of October 25-30, 1999, culminating with a six-hour educational fair at an event space on Rensselaer's campus.

I assembled six colleagues to organise the program and attained support from interim provost Doyle Daves for acquiring and covering the cost of the facilities. We prepared 300 packets containing information related to Y6B and promotional materials for the educational fair and sent them to public and private schools, local community governments, and organisations in Rensselaer County. Just under an inch thick, the parcel contained 26 items, including materials provided by Zero Population Growth (ZPG). When I called ZPG to request the handouts, the man on the phone asked, "Did your daughter, Virginia, go to Grinnell?"

"Yes," I responded, to which he replied,

"I was at the population talk you gave my first year at Grinnell. That talk is the reason I am at ZPG." Educators—people who facilitate learning—rarely know which seeds they had sown actually germinate.

The educational fair featured 18 films, including *America's Endangered Species: Don't Say Good-Bye*; *Ancient Futures: Learning from Ladakh*; and *Subdivide and Conquer: A Modern Western,* screened throughout the day. Nine workshops covered topics ranging from Big Numbers and Exponential Growth to Culture to Sprawl. Rensselaer's Great Hall in the Communications Building, almost the length of a football field, was brimming with tables and display boards from churches, departments, environmental groups, government programs, educational organisations, and schools. At one entrance we placed a continuously repeating, seven-minute film titled, *World Population* on the history of world population growth, illustrated with dots that represented one million people. Many people stopped to view the eye-catching film.

The organising committee promoted the education fair widely, especially at their institutions. One of my experiences is worth noting, because it demonstrates the aversion some scientists have to stepping into the public domain—and then shunning those who do. A prime example is the denial of membership in the U.S. National Academy of Science to Carl Sagan because he promoted space on television. At a biology faculty meeting where I was seeking support for the event, the department chair and several others chastened me for 'advocating' population issues. We received no departmental support, and but one biology colleague among a dozen stepped forward to help. He designed an attractive cover for the educational fair brochure, helped at the event, and was one of four science faculty members who attended—all were personal friends.

The fair itself was a resounding success. It brought together representatives from 22 Rensselaer County organisations, all of which had recognised a connection with the Y6B challenge and promoted their agendas to 150 attendees. I was affirmed and pleased by the hundreds of hours put forth by a few dozen people to make our educational outreach efforts impactful.

*****

None of the persons presenting or working at the educational fair anticipated financial compensation for their time. After all, this is what is expected of educators—we, by and large, provide our expertise as a community service.

When I was new to Rensselaer, the College Entrance Examining Board and the Educational Testing Service were preparing a first-year college biology equivalency examination. To ensure the quality of the questions, the exam was pre-tested. My department chair asked if I would be interested in overseeing 150 pre-college students as they took the exam. On my response form to the College Entrance Board, I wrote two questions: Is there financial compensation to the biology department for faculty and student time spent on this testing? And if not, why should we devote our time to this project? The answers were, "there is no financial compensation… benefits are intangible and, in the future, … Better examinations mean better entering students."

Many years later, when I was director of the environmental science program, a Rensselaer trustee who was a vice president of a major company asked if I would assess a new environmental biology program at Sussex Community College in northern New Jersey that his wife had developed. The assessment was required by the state of New Jersey. The trustee offered me $250 for a day on campus and for two days of review writing. I thanked him for the offer, then asked him what his company paid for a professional consultant for three days. Taken off guard, he eventually replied, "Perhaps a $1,000 per day." I replied, "Can you reframe your request by explaining that the school cannot pay me for the assessment, but could provide a token $250 honorarium?" He did. Our son, Stuart, who was home preparing to go to graduate school in biology, came with me to help and learn.

I was not surprised by the trustee's initial offer because it indicated the low esteem in which some Rensselaer trustees held faculty members. Other interactions I had had with Rensselaer trustees, especially at the annual faculty-trustee dinner, indicated an inability to accept faculty members' values. Two trustees sat at the table with me and half a dozen other faculty members. In conversation, the topic of salaries came up. One of the trustees commented that when deciding upon which job to take, a person always goes for the highest pay. I responded, "A year or so ago I considered going to a new biotech company that offered twice my Rensselaer salary, but decided that I preferred the academic life." As if he could not comprehend my remarks, the second trustee repeated what the first had said: money is the deciding factor. A new chemistry faculty member chimed in, "I came to Rensselaer instead of industry where I had offers twice my Rensselaer salary because I wanted academic freedom and to teach." Unbelievably, the first trustee repeated his initial remark. I looked at my chemistry colleague and she at me, trying to unpack what had just transpired. At the time, essentially all Rensselaer trustees were businessmen—academics, for the most part, were cognitive aliens to many of them—and deep down, I suspected most believed our jobs were easy and thought we were overpaid.

*****

When David Orr and Keith Harrison in 1994 came to Rensselaer to brainstorm ideas for our One Mile of the Hudson River course for an evening and a day, we provided each a $1,000 honorarium plus expenses. Orr was so taken by the course that he taught a class on the Black River at Oberlin the next year—Oberlin is located in its watershed. Orr had joined the Oberlin faculty in 1990 to head up its emerging environmental studies program. In 1979, he and his brother had founded Meadowcreek, an environmental education centre in Fox, Arkansas. I began interacting with Orr soon after his arrival at Oberlin and quickly realised that he was a visionary of the first order. Oberlin hired him to grow its Environmental Studies Program and grow it he did! Within 10 years, the program was the third largest degree program at Oberlin College, graduating some 40 students annually. As director of Rensselaer's undergraduate environmental science program, I knew that Orr's thinking was years ahead of most everybody's.

I wrote a letter to the Oberlin College trustees and administrators, urging them to champion the Environmental Studies Centre that Orr had proposed. Not only was it a fantastic idea, but also Orr had offered personally to raise money for construction and an endowment for maintaining the centre. Three people responded. The college president stated clearly that this was not how the college decided major commitments. The board chair agreed with the just-finished Strategic Issues Study that had recommended Environmental Studies as a high priority for Oberlin. Another trustee appreciated having his attention focused on curricular concerns and the centre.

When facilitating learning with the hopes of promoting change, a person never knows if change will happen or what will eventually emerge from the interaction. Neither trustees nor elements of the college championed Orr's vision for a centre that would teach ecological literacy to all students. Rather it was Orr's own unwavering commitment to architecture as pedagogy and to maintaining a decent planet for the future that kept the idea alive. The result was the Adam Joseph Lewis Centre for Environmental Studies, which opened in 2000 as one of the most advanced examples of green building in the United States. Orr's writings, talks, and actions were central to catalysing the environmental movement on campuses nationwide. All the while, as Orr overcame one obstacle after another, I asked if I, as an Oberlin alumnus, could help. He always answered, "No, not at this time."

Various colleges and universities longed to have Orr at their institutions, and some made efforts to recruit him. An Oberlin colleague and botanist, David Benzing, and I grew concerned that Orr would tire of the perpetual resistance to his initiatives and jump ship. A few times it seemed likely, so we reminded him of important accomplishments that he had made and that he would have to start from square one at another campus. I became more concerned once the Lewis Centre was up and running—Orr's off-campus activities were claiming his interest. In the fall of 2001, without consulting Orr, I moved forward in creating an Oberlin alumni affiliate group called EnviroAlums to advocate for environmental engagement at Oberlin and beyond. EnviroAlums's stated purpose was:

As alumni/ae who have concerns about environmental issues like biodiversity loss, climate change, consumption, equity, pollution, population, and social justice, we have formed an alumni affiliate group to assist with, and to advocate for, environmental education within the Oberlin community. Although the nature and extent of this assistance and advocacy will emerge as the EnviroAlums becomes functional, we envision providing advisory input, encouraging and supporting environmental educational venues on and off campus, encouraging and nurturing support from all elements of the Oberlin community, and raising funds for environmental education.

I put together a founding committee of eight alumni representing six class years that ranged from 1936 to 1994. In March 2002 EnviroAlums (E) sent a letter to all environmental studies graduates and other alumni supporters. The response was encouraging: 110 alums joined E, and 25 expressed interests in serving on the steering committee. I invited the group of 25 to an organisational meeting in September 2002. Twelve alumni and two students attended, thereby becoming the official E steering committee. We approved the proposed bylaws and decided to plan a three-day campus event for the following June: "Climate Justice: What is Oberlin College's Role in Climate Change?"

At that event, Amory Lovins, CEO of the Rocky Mountain Institute and a world-renowned energy expert, gave an inspirational opening address that explored Oberlin's challenge of becoming climate neutral. The audience of 27 was uniformly impressed with the presentations and the energy of the meeting. Following the event, attendees sent letters to Oberlin's president and trustees supporting the environmental policy under consideration. At its March 2004 meeting, the board of trustees approved the most comprehensive, visionary environmental policy adopted by any college or university in the United States. And while the policy set a goal of climate neutrality for Oberlin, no money was allocated for its implementation.

Despite positive signs of forward movement by the college, E's fall 2004 steering committee meeting was disappointing, attended by just two steering committee members and me, plus Orr, Benzing, and two other guests. Benzing, a member of the biology faculty since 1965, had been associated with environmental programs at Oberlin since their beginnings. He served on the Environmental Policy Advisory Committee—chaired by Orr—during its five-year existence and edited its report that the trustees approved in March 2004.

Benzing recounted the history of the Environmental Studies Program, noting that Oberlin's president had granted Orr permission to raise funds for the Lewis Centre under three conditions: he would have no assistance from the development office and he could obtain no funding from anyone who had donated or was likely to donate to the college nor any foundation that funded college buildings. Orr was given two years. Just three months shy of two years, however, the president told Orr she planned to cancel the project at the next board meeting because she did not believe he could raise the funds. The next day, Orr arranged a meeting with his best prospect from whom he secured the naming gift of $3.2 million.

Benzing went on to describe how the Lewis Center had become a beacon that attracted students and inspired off-campus initiatives in academia and in public, private, for-profit, and not-for-profit settings. But at Oberlin, efforts to green the campus continued to be a real struggle. Truth be told, sustainability advances were due only to the persistent efforts of some half dozen dedicated people.

Benzing believed that the college administration had failed in its once-in-a-century opportunity to affirm Oberlin's historical reputation of doing the right thing at the right time, despite strong headwinds. Oberlin squandered a unique set of circumstances in the 1990s that positioned it to lead the green campus movement well into the 21st century. In a word, Oberlin lacked vision. Effectively addressing environmental challenges is what the 21st century needs to accomplish—arrest climate change, biodiversity loss, population and consumption growth; clean up toxins while preventing release of more; transit from industrial agriculture to ecologically grounded food systems—to name a few. That is, our economically centred culture will have to become ecologically centred if humanity is to achieve a durable future.

Benzing's assessment was affirmed by *Sierra Magazine's* rankings of 'the coolest schools', meaning the greenest. In early years Oberlin consistently ranked in the top 20—even topping the list in 2007 when first established. In 2017, it fell to 19. Then, Oberlin went off the top-20 list entirely.

Despite Benzing's perspective—which I accepted—E had to carryon. I solved the lack of attendance at committee meetings by setting up conference calls for those who could not attend in person thereby increasing participation to nearly 100 percent. As with most volunteer organisations, many ideas were proposed, yet volunteers are few. Much of the work fell to me, as chair. I adopted a new practice of obtaining consensus when projects were proposed, and only if someone stepped up to lead the project would it go forward. Finally, thanks to committed volunteers, dues, and donations, E's outreach expanded. We brought alumni to campus for career talks; awarded small grants of several hundred dollars for student projects; donated $1,000 to the environmental studies program for career development programs; funded a program by the bicycle co-op to provide bikes to alumni returning for reunion weekend; distributed copies of Tim

Flannery's book, *The Weather Makers: How Man Is Changing the Climate and What It Means for Life on Earth* to administrators, staff, and faculty to promote campus discussions; and sent DVDs of Al Gore's *An Inconvenient Truth* to trustees so they could see reasons to support the college's environmental policy.

I love endowments, and E needed one to do more. We organised a two-step process. The steering committee held a mock Climate Justice Challenge Campaign (CJCC) in November 2006 to gauge the potential level of financial support if Oberlin were to place environmental sustainability front and centre in an upcoming capital fund-raising campaign and in its search for a new president. E's Steering Committee pledged $35,905, and 27 alumni pledged $22,075, for a total of $57,980. The CJCC statement and pledge results were provided to the president and trustees at their December board meeting.

Neither the college president nor trustees acknowledged the mock CJCC. Nevertheless, it laid the foundation for E's steering committee in December 2007 to enthusiastically and unanimously move forward in establishing an endowment. The executive director of the alumni association, the vice president for development, and the college president approved the E Endowment Fund in February 2008. In that same month, E steering committee members unanimously committed to making five-year pledges to the E Endowment Support Fund and asked all E members to do the same. By June 30, 2008, we had raised $72,419, enough for an endowment, plus an additional $35,690 in pledges. I was astounded! I was led to believe by the director of the alumni association and the vice president for development to expect several years to raise the $50,000 needed to make the fund an endowment.

Oberlin has a grand tradition of leading the way toward a fairer, just world. E acted on the belief that this could be one of Oberlin's finest hours, if together we could seize the opportunity through visionary environmental leadership. E also needed knowledge of campus activities and concerns to effectively promote Oberlin's environmental agenda. From the beginning, John Petersen, an Environmental Studies faculty member and Oberlin graduate, had been our campus representative on the committee and had recruited one or two student representatives each year. Some went on to become steering committee members after graduation, for a three-year term. We often elected one as vice chair so as to nurture a few who could take charge when the time came.

I remained chair for a dozen years. E was successfully engaged on and off campus—organising symposia, meeting with alumni clubs, serving on the alumni council executive board so as to bring an environmental perspective to alumni leaders, organising panels on environmental topics during reunion weekend, providing books and materials to faculty members and trustees, and volunteering to host students for January-term and summer internships. By 2014, it was time for the younger generation to take charge. Two recent graduates who had been active in the group became chair and vice chair. When my fourth term on the steering committee ended, I made my exit, emboldening the group to do what *they* thought best. I walked away to seek other opportunities, knowing that E had met a significant challenge.

*****

For most of my life, I have taken photographs, initially print images with a box camera and then slides with 35 mm cameras. Long before the age of PowerPoint, I used photos in my talks and class lectures to visually present subject matter. The tragedy in Nauru, for example, I conveyed with a picture depicting Nauru's native tomano tree (*Callophyllum inophyllum*) ecosystem in the background and naked coral pinnacles interspersed with phosphate ore in the foreground.

7.1: Phosphate mining on Nauru, native tomano tree (Callophyllum inophyllum) ecosystem is in background with foreground of mined out area showing remnant coral pinnacles from between which phosphate ore had been taken.

Video, too, was perfect for capturing the dire consequences of Nauru's adoption of a capitalistic market system. Mary and I spent a month there trying to obtain permission to record a documentary. However, Nauruans had had unpleasant experiences with westerners and had no desire to allow it. I did take many pictures and some video, which led to a 15-minute video that I shared with colleagues for use in their classes. This was a visually engaging example of how our global market economy was affecting the world.

In my early years as a scientist, I focused most of my professional talks on our science. Over the past 20 years, that focus has shifted to community service.

My engagement with the Science Teachers Association of New York State (STANYS) gives flavour to that aspect of my community service.

STANYS, founded in 1896, is the oldest professional organisation of science educators in New York. The northeast section of STANYS held their annual meeting in the fall at Siena College, just across the Hudson River from Rensselaer. I began attending their annual meetings in 2000, when the Union of Concerned Scientists and the Ecological Society of America began a Communicating Ecosystem Services Project and provided tool kits for water purification and pollination. I had been teaching aspects of ecosystem services in my introductory biology class and writing about it in my research on ecological economics with John Gowdy. These made-to-order kits were perfect in conveying to high school science teachers the importance of biodiversity loss and the life support services (i.e., ecosystem services) affected by it.

Joan Wagner, chair of the northeast section, invited me to give a talk on John Gowdy's and my book about Nauru, *Paradise for Sale*. Not long after, she recommended to STANYS' vice president that I speak at STANYS' annual conference in November 2002. In my keynote address, titled *Hopefulness in a Time of Peril,* I talked in detail about evolution as the uniting concept in biology. I went on to explain that evolution is a law in biology in the same way that gravity is a law in physics. A number of audience members walked out. It seemed a strange response from science teachers, but on reflection I realised that many of them likely had students whose families found evolution incompatible with their religious beliefs—perhaps some high school biology teachers believed likewise. As a result, many high school biology teachers avoid teaching evolution. It is little wonder that the majority of people in the U.S. do not accept evolution as the process by which the diversity of life came into existence.

*****

I enjoyed writing and talking about *Paradise for Sale.* I had learned so much in the process about the topic and had found writing the book enjoyable that I decided to write another book: *Wisdom for a Livable Planet*. In it I told the stories of eight visionaries who had dedicated their lives to addressing what I thought were the mega environmental challenges facing humanity. Herman Daly advocated a dynamic steady-state economy respectful of the laws of nature and human behaviour. Dave Foreman sought to restore functional ecosystems and preserve biodiversity through the Wildlands Project. Werner Fornos strove tirelessly to provide every person with the wherewithal to decide when and how many children to have. Wes Jackson promoted sustainable agriculture based on local ecology and community values. Helena Norberg-Hodge fostered local living through appropriate technologies. David Orr championed educational reform, thereby enabling students to become environmental-aware, proactive citizens. Stephen Schneider addressed the challenge of climate change, and Terri Swearingen employed activism to curtail the release of toxic waste into the environment. The book's final chapter was titled, *Can We Change, Will We Change?*

Writing *Wisdom for a Livable Planet* solidified and focused the following understandings in my mind: evolution can only select for behaviours that are successful in the present. For example, the *hominin* brain evolved to fit hunter-gatherer societies; our evolved behavioural repertoire is that of a small-group, place-based social animal that is territorial and hierarchical. Anatomically, modern *Homo sapiens* evolved in and then migrated out of Africa some 100,000 years ago into ecosystems that appeared limitless. Our behaviours have evolved in compliance with this boundless frontier.

Our ancestors acquired language and became storytellers. They sensed their immediate surroundings but knew little of why and how the biological-physical world worked the way it does. Picture a tobacco plant with its terminal flower in bloom and answer this question: "How does a plant make roots, shoots, leaves, and flowers?" using **only** your evolved senses and the unaided power of your brain to venture an explanation.

Take a few minutes to articulate your explanation.

With confidence, I know that any answer you venture will be wrong. Why? The micro-world of cells, genes, and chemical signalling is hidden from us. It is beyond our unassisted abilities to discern a world so removed from human experience.

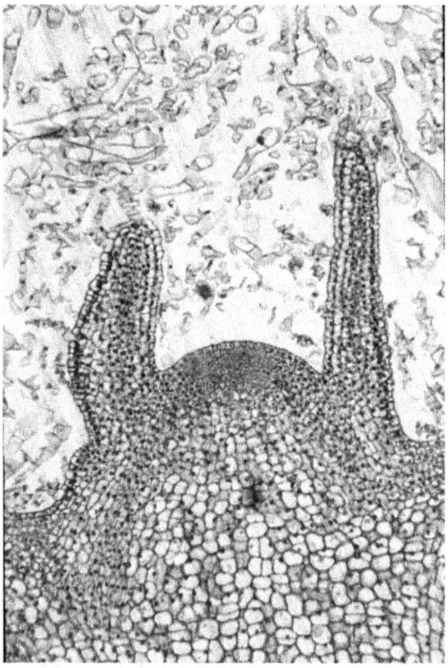

7.2: A microscopic longitudinal section of a vegetative shoot apical meristem of tobacco.

To explain the world about them, our ancestors created stories. They believed the stories. Some of these imagined relationships were adaptive and enabled survival. In time, the act of believing became part of human nature. What does this mean? Beliefs that are deeply held have permitted a culture that denies limits.

Why is this important today? What a culture teaches and what people believe may be false, and dangerously so when important decisions are based on such beliefs. The western cultural belief in unlimited growth ignores a biological and physical reality, thereby hampering any resolution of our environmental challenges.

Important environmental trends are almost wholly negative—rapid global warming that drives ever more extreme weather events, an increasing and significant loss of biodiversity and the life support it provides, an increasing use of toxic substances and novel materials like plastics, the loss of arable land, and new and spreading diseases. Yet the growth of human activities—the underlying cause of these trends—has been advocated. To act on the consequences of these trends is extremely hard for an animal that has evolved to act in the present. Edward O. Wilson encapsulates the conundrum:

The human brain evidently evolved to commit itself emotionally only to a small piece of geography, a limited band of kinsmen, and two or three generations into the future. To look neither far ahead nor far afield is elemental in a Darwinian sense. We are innately inclined to ignore any distant possibility not yet requiring examination... Why do [humans] think in this short-sighted way? The reason is simple: it is a hardwired part of our Palaeolithic heritage. For hundreds of millennia those who worked for short-term gain within a small circle of relatives and friends lived longer and left more offspring—even when their collective striving caused their chiefdoms and empires to crumble around them.

This hardwired behaviour notwithstanding, cultural change is required to avoid civilisation collapse on a global scale similar to that routinely visited on cultures locally when an element(s) of life support played out—Easter Island (Rapa Nui), Mangaia, Mangareva, Henderson, and Picairn in Polynesia; Mayan and Anasazi in Americas; Maltese in Mediterranean; Sumer, Egypt, Mycenae, and Petra in Middle East; and Harappan in India's Indus valley.

Although most local groups came undone from the loss of life support, a few avoided that fate. Kalahari !Kung in Africa apparently typified those hunter-gatherers who lived well, unconstrained by their environments. The Nauruan culture existed isolated for thousands of years without impoverishing its important resources. Ladakh, situated in northern India on the Pakistan border, harboured a durable culture until the 1970s when its isolation was breached. Tikopia, a small two-square-mile island in the Pacific, was ravaged for 800 years. The Tikopians, recognising their resource losses, ceased their population growth and restored much of their biodiversity. These durable cultures give promise because they establish that humans have the capacity to live within their resource limits and, in the case of Tikopians, to significantly change their cultural behaviours so as to arrest collapse and become durable.

The ultimatum now before our current, growth-based, global culture is to affect the transition from an economically centred culture to an ecologically centred one. Unfortunately, this shift will require monumental changes in cultural beliefs. For example, European and North American nations have economic systems that require continual growth; yet, continual growth of material things is not possible in the finite system that earth is. These economic systems have been constructed based on the belief that the natural world from which resources are taken is a part of human economy; yet, the converse is true—human economic activity occurs within the natural environment. Coupled with this belief is the belief that earth's bounty is here for human use. Fundamental beliefs are glacially slow to change, if they change at all.

The history of the Catholic Church and its acceptance of evolution are illustrative of the acceptance of a new reality that replaces a no longer tenable belief. Darwinian evolution based on the 1859 book, *On the Origin of Species*, attained scientific consensus by 1900. In 1996, Pope John Paul II accepted evolution. However, most people in the U.S. reject evolution by natural selection as the means by which Earth's diversity of life came into being, especially humans. Time is not on our side to transit to an ecologically based culture before our economically centred one collapses civilisation.

Chapter 8
Science in Retirement

I walked between two patches of woods and part way down a gentle slope. The autumn sun warmed me as I sat and then lay back on the grass. An occasional bird flew through my vista of blue sky in this enchanting scene, contrasting with the leaves beginning to reveal their autumn colours.

I lay in the centre of five biologically diverse acres located less than a mile from Tappan Square, the 13-acre green space connecting Oberlin College to the centre of the town of Oberlin. Just shy of nine months earlier, Mary and I had decided to abandon the place where we had lived for most of our adult lives and strike out on our final phase together: retirement. For decades, we had coveted Hill Top, David and Harriet Borton's passive-active solar home in Troy that ran on sunshine. Now, we owned these five acres of land in Oberlin—a perfect site for us to replicate Hill Top.

In December 2006, we had decided that Oberlin provided unique opportunities for our retirement. Mary's sister, two brothers, parents, and grandfather had graduated from Oberlin College, as did our son and me, making this a familiar place. I had been collaborating with David Orr since 1990 and was very active in alumni affairs, especially EnviroAlums. Orr had expanded his visionary horizon from 21st-century sustainable buildings such as the Lewis Center to sustainable communities. Orr believed the City of Oberlin was small enough to accomplish significant change, yet large enough to serve as a replicable model.

To act on this proposition, Orr initiated the Oberlin Project in 2009. I had been interacting with Orr for almost two decades and a major motivation for coming to Oberlin was to join the effort to make the community durable. Over the Oberlin Project's eight-year life it was the driving force that brought the City, College, and townspeople together in common understanding that climate change must be addressed as well as resilience and inequities. I along with staff of the Oberlin Project played significant roles in the writing and adoption of the City's Climate Action Plan in 2013. However, our major contribution to the Oberlin Project was the building of our home Trail Magic.

Don Watson, a friend and former dean of architecture at Rensselaer, designed Trail Magic to acquire solar energy passively and actively. We were bringing our 3.1 kW solar system from Troy to provide our electricity. But would it? Because Trail Magic was our gesture to Orr's vision of a sustainable community, I knew well that achieving this goal in the first year was critical—the initial success or failure is what people remember.

I had employed a Kill-a-Watt meter in Troy to estimate our plug load—the electrical draw of all 120-volt devices that are plugged into the electrical system of a house. When the Kill-a-Watt meter is plugged into an outlet and a device into the meter, it displays electricity flow (Watts) and the amount that has flowed (kWh) over the time period being measured. We could control plug-load, as well as hot water use and mechanical heating and cooling. I measured the Watt draw of our devices: Internet connection: 14; laptop computer when running or at standby: 20 and 1, respectively; TV on or at standby: 75 and 1, respectively; kitchen radio-CD player: 1 to 4 depending on volume; stereo system and CD player: 30 to 100 depending on volume, and more. This information enabled us to conserve electricity: for instance, we turned off the Internet when not in use and often used the kitchen CD player instead of the stereo system. Yes, 14 watts for the Internet connection was not much, but a year's use equalled 122 kWh—and that could impact the ability of our solar system to meet all electricity use.

Mary and I made many other decisions that would conserve electricity without affecting our quality of life. We installed compact fluorescent and LED bulbs, Energy Star appliances, and low-flow showerheads (1.5 gallons per minute). We dried clothes on a clothesline, took short showers, and turned off lights and appliances when not in a room. The moment of truth came on November 12, 2009, one year after moving in—we had used 2,279 kWh and made 3,192 kWh! Over the next four years, our use and production averaged 2,900 kWh and 3,200 kWh, respectively. Our home was powered by onsite sunshine.

8.1: Trail Magic south and west sides with 2-year tall grass prairie in the foreground.

Although Trail Magic was similar to other houses across the country in its capacity to run on sunshine, as a scientist I documented everything, beginning with designing and building Trail Magic up through its performance over the first four years. This led to a notebook full of experiments, as well as a book, *Trail Magic: Creating a Positive Energy Home*. Watson noted, "Your scientific data are what makes Trail Magic truly unique."

The team who designed and built Trail Magic believed that its most advanced feature was an evacuated-tube solar hot water system that cost $6,000. Solid copper cores housed in glass evacuated-tubes on the roof captured solar heat that was transferred to a glycol solution. The heated glycol was pumped through a heat exchanger that preheated water in an 80-gallon tank (a standard electric hot water heater not wired for use). When hot water was needed, the preheated water flowed through an on-demand hot water heater that warmed the water to the desired temperature, if needed. A really slick system! But I wondered, "How much heat in the hot water that we actually used came from the sun?"

The man who sold us the evacuated tube system was also interested in the answer. He installed a water meter on the preheat tank to measure our hot water use. I measured the amount of electrical energy needed to pump the glycol and water and the amount needed by the on-demand heater. Thanks to a thermometer in the pre-heat tank, I quickly discovered that most of the heat acquired during a sunny day was lost overnight because the preheat water tank was poorly insulated. So, we tripled the insulation, which reduced heat loss by half.

I measured the amount of electricity needed to pump fluids and used by the on-demand heater along with temperature of incoming water, temperature of hot water, and gallons of hot water produced to determine the actual amount of solar heat in the hot water we used. The solar energy in the hot water used for one year was about 96,000 British Thermal Units (BTUs), or $2.80 at $0.10/kWh. Wow! What was the problem? The evacuated tube system was very efficient and acquired nine million BTUs annually. However, we only used annually less than one percent of those BTUs for heating hot water.

Unfortunately, few people know the number of gallons of hot water they use, nor can they easily determine the energy employed to heat their water unless they get the relevant numbers, as I did. David Borton, who built a home that runs on sunshine and was my best friend in Troy, NY, and I talked about the evacuated-tube hot water system but did not think to ballpark how much energy might be used if we heated with an on-demand hot water system. Our annual use of 2,500 gallons of hot water is now met by an on-demand system for around $40 at $0.10 per kWh, if we purchased electricity.

The man who sold us the evacuated tube system offered to rebate us half of its cost if we put the money toward a second PV system. We did. The new 2.1 kW solar system gave Trail Magic a total of 5.2 kW that produces annually about 5,200 kWh. These additional kWh gave us the capacity to cover increased demand. In 2012, we replaced our 1995 Saturn with a Toyota Plug-in Prius that provides with our solar electricity about 25 percent of our annual mileage consuming around 750 kWh. We have subsequently replaced that Prius with a

2018 Plug-in Prius that gives twice the miles per charge thereby using annually about 1,500 kWh.

*****

Although it is extremely difficult to accurately measure solar heat gain in a house because of so many uncontrollable variables, the heat gain in ours is considerable. On January 16, 2009, at 10 a.m., the temperature was 62 degrees inside and minus 10 degrees outside. The sun was bright in a clear blue sky, and by 2 p.m. the temperature had risen to 68 degrees inside and 0 degrees outside, an inside gain of 1.6 degrees per hour. These and other data indicated that full sunshine provides enough energy not only to offset heat loss, but also to warm the house when the temperature differential is about 65 degrees.

We also measured the heat loss at night. On January 23, 2009, at 9 p.m. the temperature was 70 degrees inside and 6 degrees outside. The night was calm, and by 8 a.m. the next day, the temperatures were 62 degrees and 1 degree—a house loss of 0.7 degrees per hour. An earlier night measurement had verified that Trail Magic was so tight that wind had little effect on heat loss. On December 21, 2008, at 9 p.m. the temperatures were 70 and 0 degrees. The wind blew all night at a constant 20 to 30 mph. At 8 a.m. the next morning, the temperatures were 61 and 5 degrees, for a house temperature loss of 0.8 degrees per hour—just 0.1 of a degree more than on a windless night.

A blower door test established about 1.1 air exchanges per hour at minus 50 pascals (measurement of reduced pressure [vacuum]). To create this vacuum, an entry door of the house is removed and replaced with a door that has a fan in it that blows air out of the house to create a vacuum. The typical house would have between four and 10 air exchanges per hour at minus 50 pascals.

When compared to a standard U.S. home in terms of energy used and purchased, Trail Magic stacks up well. The average home buys and uses annually about 100 million BTUs, while Trail Magic runs on 35 million BTUs: five million from passive daylighting, 20 million from passive heating, eight million for heating and cooling with a heat pump, and seven million for all the rest (lighting, hot water, electronics, and appliances). Trail Magic uses 15 million BTUs of electricity directly from solar that is its primary energy. Fossil fuels—coal and natural gas—are burned as the primary energy source to produce the electricity that most people use in their homes. To produce each BTU of electricity it takes more than three BTUs of fossil fuels because fossil fuels are mined, transported, and burned to produce most grid electricity. When all of the above are factored in, Trail Magic uses less than five percent of the primary energy used by similar homes in Oberlin: 15 million BTUs (electricity) ÷ (100 million BTUs x 3 BTUs/1BTU) x 100 = 5 percent.

OK, but what did it cost to build Trail Magic? The actual cost was $146 per square foot for 2,494 square feet of inside space, which is the typical square-foot cost of a modest custom-built house in northeast Ohio.

Our builder estimated $91,000 for custom features that were not required for the home's performance (such as a 24-gauge metal standing-seam roof for $17,000, which could be exchanged for a 25-year asphalt shingled roof that cost $5,500) lowering the cost to $110 per square foot, or the price of a high-performance tract house. Casey Gilfether, who had built hundreds of tract houses, became interested in Trail Magic near the time it was finished and read my book on Trail Magic four years later. He was impressed with the house and found the book informative but thought my estimated square-foot-cost for a high-performance tract house was too low.

*****

In the spring of 2013, David Orr, head of the Oberlin Project, and David Sonner, chair of Zion Community Development Corporation's board of directors and the real estate person who had found our property, were each contemplating building an unpretentious, high-performance house in Oberlin to demonstrate that high performance homes are affordable for people of modest means. Both asked me to be involved. By late fall, it became apparent that building two houses would be impossible, as even one house would be a huge undertaking. They agreed and decided to collaborate. This would be the first house in Zion CDC's affordable green housing initiative.

After locating a perfect lot on Smith Street, I agreed to coordinate a project team of Sonner, Gilfether, and Orr along with architects Joe Ferut and Lee Goodman who did the detailed design work for Trail Magic and Alan Mitchell from Oberlin Community Services, who vetted applicants and was the project's fiscal agent. Gilfether became the general contractor and the building-talent that made the project successful. He and I knew this would be an experiment to test my hypothesis that a high-performance house—defined as using less than half the energy of an equivalent house in Oberlin—could be built for the typical cost of a northeast Ohio tract house.

The Smith Street house team had several major hurdles to clear. First, we needed a buyer who could qualify for a USDA Rural Development Mortgage—someone who would not qualify for a regular mortgage, but whom the USDA would consider capable of managing one. Second, we needed to secure a variance to an outdated Oberlin City ordinance that required the first floor of a house be larger than 1,000 square feet. Third, we needed to secure a $75,000 bridge loan to cover initial construction costs not yet provided by the USDA—the agency held 40 percent of its payment until the home was occupied. Fourth, we had to keep the project moving while Gilfether recovered from quadruple bypass heart surgery. Amazingly, despite the fact that the Smith Street house team had never worked together, we cleared all the hurdles to deliver a high-performance home to the perfect client.

The client (a woman with two teenage daughters) demonstrated behaviours that were consistent with attaining a high-performance home. In the first year, they used 22 million BTUs compared with 104 million BTUs used by similar homes in Oberlin. The 1,409-square-foot house cost $115 per square foot. Gilfether accepted my hypothesis that a high-performance, quality tract house could be built for about $110 per square foot in northeast Ohio.

I need to note, however, that the behaviour of a home's occupants is critical in creating a high-performance or positive-energy house—anybody can waste energy. In fact, experiments have shown that energy use varies by a factor of two in very similar houses in the same location and occupied by an equal number of people. These data indicate that homebuilders can design and construct houses that are *capable* of being positive energy, but not guaranteed.

*****

Clearly, building houses with the capacity to run on sunshine is a necessary part of creating durable communities—but that alone is not enough. Landscapes designed for the 21st century, among other things, are also necessary. With Trail Magic's five acres of land, we strove to accomplish six goals: keep precipitation on site, enhance biodiversity, grow some of our food, sequester carbon, eliminate invasive alien plant species, and make the landscape basically self-managing.

Oberlin was founded in swampy woods of pin oak and elm trees and proved ill-suited for agriculture. The five acres we purchased were among the last of the best land in Oberlin, a distinction of questionable significance in light of Oberlin's location. The founders of the city and college were seeking an isolated place absent of people and several miles away from populated areas—land shunned by others—to establish their Christian colony and college. They found land in the middle of what is now New Russia Township and in 1833, some 20 acres were cleared of trees for colony and college.

Trail Magic's landscape is a five-acre rectangle bordered to the north by East College Street and extending south to the Plum Creek floodplain; the southeast corner ends in the creek. The land slopes gently south with a 22-foot drop and contains a diversity of ecosystems—two grassy meadows and patches of woods with trees of various ages and sizes. These include mature red oaks that measure 75 feet tall and 3 feet in diameter, sugar maples, elms, Osage orange trees of various sizes, native dogwoods, wild crab apples, and originally some three-dozen mature ash trees that were fated to die from the emerald ash borer. Five years after lumbering our ash trees, no ash trees of any size remained alive in Oberlin. We lumbered ash trees, a few black walnuts, two dying red oaks, and a sick red maple for our flooring, shelves, bookcases, trim, beams, counter, and dining room table. Invasive species—buckthorn, Japanese honey suckle, privet, and Japanese knotweed—were present, but have mostly been removed.

*****

The soil is fine glacial clay with minimal percolation, meaning if you dig a hole and fill it with water, the water will remain in the hole for days. Little percolation is ideal for creating a pond—just dig the site, and it will fill with rainwater. When dry, the clay is hard; when wet, it is slicker than a greased pig on the fourth of July and sticks tightly to boots, clothes, and skin. During construction, I out-and-out hated Oberlin mud.

The builder of Trail Magic did everything possible to prevent water from seeping into our house. Almost all homes in Oberlin have water in their basements following a heavy rain. A pond was dug for fill to raise the house site some six feet, but also to store rainwater, raise fish, provide a heat source/sink for the heat pump to cool and heat the house, and for aesthetics.

Thirty years ago, most homeowners managed water on their property by getting it onto a neighbour's land as fast as possible. The converse is true today—precipitation should be kept on site as long as possible so as to prevent flooding downstream. We designed the landscape to control flooding by constructing swales to drain water around the house and then into the pond or down the grassy meadows to the flood plain.

We began restoring the flood plain soon after removing the treetops from lumbering. Similar to many out-of-sight flood plains, ours had attracted all manner of discarded items: concrete slabs, asphalt, pieces of sandstone, car bumpers, steering wheels, baby carriages, scrap metal, and box springs, to name a few. We hired the excavator who dug our pond to bury the debris and grade the flood plain to make it accessible to my small tractor for mowing and maintaining brush. We then constructed a dam across the trench through which storm water had been flowing into Plum Creek, thereby containing onsite a 24-hour, 3-inch rain.

*****

Enhancing the biodiversity has been ongoing. We removed mechanically a large fraction of the Japanese knotweed that was in our portion of the flood plain and then spread seeds of flood-plain plants. We continue to remove other invasive plants throughout the property: buckthorn, privet, Japanese honeysuckle, and garlic mustard. In the restored floodplain, in the swale between grass meadows, and in the front yard, we planted some 200 native tree and shrub species, some of which are now over 30 feet tall.

About an acre and a half of tall grass prairie was planted in the backyard, the northern quarter of a meadow, around the pond and garden, and in the swale between meadows. The backyard area was a test plot that is now nine years old with big blue stem, switch grass, and Indian grass being the dominant species. The rest of the prairie is four years old, with the tall grasses just starting to show. These tall grass prairies are similar to what produced the rich soils of Indiana, Illinois, and Iowa following the retreat of glaciers 10,000 years ago, because of which they have become among the best farming land in the world.

Tall grasses put down roots six feet deep, and some 65 percent of their photosynthetic product is put into roots, thereby sequestering more carbon than most trees and other plants. Since planting the prairie in the backyard, the flow of water down the eastern swale has diminished to nothing after a downpour of a few inches. In addition to creating soil and absorbing water, tall grass prairie is habitat for innumerable insects, small birds, and mammals. It's also self-managing, aside from needing burning or mowing every several years to keep shrubs and trees from taking hold.

The pond provides an aquatic ecosystem for insects, fish, turtles, wetland plants, and algae as well as storage for precipitation and a heat source/sink for our heat pump to heat and cool the house. Our 6,000-square-foot garden has provided some of our food and in its best year—2012—produced 2,000 pounds of produce. That translates to 400,000 calories, or three to four months of caloric needs for Mary and me. In addition, the five inches of fertile soil we've made over a decade has provided habitat for innumerable soil invertebrates, fungi, and bacteria as well as sequestered hundreds of pounds of carbon.

8.2: Trail Magic garden in 2012.

*****

Most people in Oberlin and northeast Ohio had not imagined it possible to power a local home on sunshine thereby making Trail Magic an educational tool. The day before we moved-in, we had our first open house to which 350 people came. Don Watson came for the event and put signage throughout explaining the various design features and other aspects of the house that made it possible to run on sunshine, conserve resources, and be durable thereby requiring minimal maintenance: long axis east-west for passive and active solar (3.1 kW solar system); super insulated slab (R-20), walls (R-46), and roof (R-63); as small as possible but still meet our needs; triple paned custom made Loewen wood frame windows on north, west, and east sides and double paned on south side tuned for solar gain; window placement for passive heating and day-lightning; raised seam 24 gauge metal roof; foundation Reddi-Wall, framing double wall, pre-coloured

Hardiplank siding; 1,800 gallon cistern; central vacuum system; HVAC with pond loop source heat pump; hard wood floors throughout along with bookcases, closet shelves and pantry counter and shelves made from ash trees lumbered on site; Energy Star appliances; low flow showers (1.5 gal/min) and bathroom faucets (0.5 gal/min) along with dual flush toilets (0.8 and 1.2 gallons/flush).

Since that first open house, we have held a few dozen more with several thousand people touring Trail Magic. For each of these we have used signage similar to what Watson had prepared for the first open house. Some Environmental Studies courses have come annually. Joe Ferut, the detailed design and construction architect now teaching at Kent State, brings clients and students regularly. I have used these opportunities to discuss the ongoing climate change and the sixth mass extinction of life as the primary reasons for building our home to run on sunshine and the landscape to enhance biodiversity. The book I wrote on the building of Trail Magic and its performance—*Trail Magic Creating a Positive Energy Home*—has enabled others to know in detail what we accomplished thereby fostering replication in new homes and upgrading of existing homes.

*****

In 2011, David Sonner encouraged me to apply for a vacancy on the local Public Utilities Commission (PUC). Thus began my 5-year venture into Oberlin City governance. Oberlin has two utilities, Oberlin Municipal Light and Power System (OMLPS), which provides electricity, and Public Works, which provides water, sewer, road, and sidewalk services. Other than attending monthly meetings, my major contribution was helping determine the fate of about $3 million of revenue that had come from swapping high-priced renewable energy certificates (RECs) for low-priced RECs.

OMLPS purchases its electricity through American Municipal Power (AMP), which works with small community electrical utilities in offering choices of reliable electricity sources. Oberlin's primary source of base power—the flow of electricity needed 24/7—was coal-based, as it was then for most base power in the U.S. In the summer of 2007, Oberlin's city council had voted 4 to 3 to commit the city to 50 years of coal-based electricity provided by a new coal-fired plant that AMP was building. Sonner and three others ran for city council that fall on a single issue—to withdraw the city from its commitment to the coal plant. Sonner's coalition won two-year terms, and their first action in January 2008 was to cancel the contract by a 4 to 3 vote. As time passed, Sonner's supporters called the three opposing council members the "coal dust trio."

The city now had a problem! What energy source would it use for base power? As president of city council, Sonner began pushing the city manager to ask that OMLPS pursue sustainable options for power; however, the city manager and OMLPS leadership did not support the council's decision. After 18 months of inaction by the city manager, Sonner, in a public council meeting, aggressively directed him to have OMLPS identify non-fossil fuel options, earning Sonner a reprimand by the "coal dust trio." OMLPS did find landfill gas as a carbon-neutral fuel for Oberlin's base power just as Sonner's two-year term was ending.

As it turned out, the coal-powered plant was cancelled by AMP just before the city council election, leaving the cities that had signed up to receive its electricity to pay AMP the millions of dollars spent before the cancelation. Oberlin's portion would have been a little over $3 million had the city not withdrawn from the project. Sonner's thank you from voters? The coal dust trio, who had voted twice for the plant, was re-elected while Sonner was not.

*****

In spring 2007, as we began the search for a site on which to build our Oberlin home, I had inquired at OMLPS about the city's policy on net metering, which is essential when installing solar (photovoltaic electricity, PV). Ideally, the electric meter goes forward when using grid electricity and backwards when solar electricity flows onto the grid, thereby allowing use of grid electricity at night and at other times when the solar system is not making sufficient electricity to meet demand. I told the electric director that I was interested in bringing our 3.1 kW solar system from Troy and wondered if OMLPS had a net metering policy. Yes, he said, anyone could install up to 10 kW of solar. He then asked me why I was choosing solar. I explained that the burning of fossil fuels was changing the climate, and that solar was part of the solution. He ventured that climate change was not happening and that there was no practical alternative to coal. I responded that we would have to agree to disagree on climate change and solar, but that I was pleased by the net metering policy.

Oberlin, like the rest of the country, was experiencing a clash of worldviews. Simplistically, it is a clash between evidence and beliefs. As far back as 1896, Swedish scientist Svante Arrhenius had calculated that doubling our atmosphere's carbon dioxide level would lead to a temperature rise of several degrees—essentially the same conclusion made in 2007 by the International Panel on Climate Change. Even prior to 1896, basic physics established that with a lack of carbon dioxide in the atmosphere, Earth's temperature would average a few degrees below freezing. These evidence-based conclusions were contrary to the conclusions of some Oberlin residents who believed what their culture was telling them.

Bear with me as I relay a relevant incident that occurred at Rensselaer. A young woman about to take my introductory biology course asked, "I am a creationist. Do you teach biology as if evolution were true?"

"Evolution," I replied, "is the uniting concept in biology. Nothing in biology makes sense except in the light of evolution." She persisted. "Professor McDaniel, don't you know that belief is much stronger than fact?" I agreed.

Each of us has the capacity to hold and act on beliefs, even when they may be demonstrably false. Often, we are naïve, but even when exposed to overwhelming evidence that contradicts our beliefs, we are hesitant to revise them. We are a species that creates fantasies and deeply believes them. Accepting that reality, our task is to carry on with the hope that evidence and argument will win the day.

Sonner's victory in using landfill gas to generate much of Oberlin's electrical needs meant that each megawatt hour (1,000 kWh) the city purchased had one renewable energy certificate (REC) associated with it. RECs had been created in the late 1990s by states to provide an economic incentive to abandon fossil fuels for renewables in electricity generation.

The city owned its RECs, but they could be sold. With most of the city's electricity coming from landfill gas, OMLPS accrued millions of dollars in RECs. The city council took the PUC's recommendation to purchase renewable energy sources, such as solar, wind, and hydro in addition to the carbon-neutral landfill gas. The city had a goal: to become carbon neutral by 2050. Intermediate targets were also set, 75 percent carbon neutral by 2030 and 50 percent by 2015—which had already been accomplished. If the RECs were sold, the city's energy would no longer be considered 'green'. However, a robust RECs market had emerged by 2012, and the city began selling its high-priced landfill gas RECs ($14.50 per REC) and purchasing low-priced wind RECs ($2.50 per REC), thereby creating a million-dollar revenue stream.

The city now had a very enviable problem—how to use the almost $3 million in REC income. The PUC, charged with making recommendations on such matters to city council, spent numerous hours in 2014 discussing options, which centred on deciding what percentage of the funds to invest for city projects and what percentage to use in lowering electric rates. That summer I met with each person on the PUC to seek common ground on the percentages. I believed that investing the money was the best choice; for example, by re-lamping the city's streetlights with efficient LED bulbs, we could reduce electric rates for years; residents indirectly pay for street lighting through their electric bills.

Commissioner Kristen Brazunis and I, both Oberlin College graduates, were shocked by the recommendation of the electric director at our January 2015 meeting that the REC revenue be used to lower electric rates. He contended that the money belonged to ratepayers.

I looked at Brazunis and moved, "The PUC recommends that city council earmark $300,000 of REC revenue to re-lamp as many municipal street-light fixtures as possible with LED technology and hold the remaining dollars for energy efficiency projects until the PUC can come up with other possibilities."

Opposing the electric director was a bold act; he was the authority and hired by the city manager to run the utility. The chair called a vote. Brazunis and I voted yes; two others voted no. A new commissioner, Heather Elmer, also an Oberlin graduate, was participating in her first meeting. I thought, "She has been thrown into a difficult place, unfamiliar with the significance of what just took place. Would she abstain or vote?" A long, tense silence was broken with "yes." A two-year confrontation over the disposal of RECs revenue began.

Rarely does a small city, really a town of 5,000 residents and 3,000 college students, experience a windfall of $3 million, especially at a time when city revenues are declining as needs and inequities increase. Please indulge me as I chronicle Oberlin's plunge into a whirlpool of conflicting values and beliefs revealed by RECs, an economic tool designed to eliminate fossil fuel-based electricity.

In 1996, California began considering how to create its renewable energy portfolio standard. The first market to use an attribute associated with renewable electricity emerged there three years later; it was called a 'green ticket', shortly to become a REC. One REC was granted for each megawatt hour generated by renewable sources. When California instituted its renewable portfolio standard, it required the state's electric companies to continually increase the amount of green energy in their portfolios either by generating their own green energy or by purchasing RECs that other electricity producers had in their portfolios. REC markets popped up in California and then in other states that had established renewable portfolio standards.

Oberlin College sought green electricity and arranged in 2004 to purchase the RECs associated with hydroelectricity in OMLPS's portfolio. This hydroelectricity was earmarked for the college, and the revenue from the REC sale was to be used by OMLPS for sustainability projects. In 2007, city council passed an ordinance specifying the disposition of RECs revenues—an ordinance establishing a Sustainable Reserve Program (SRP) for the City of Oberlin.

This ordinance established a "Sustainable Reserve Program Account for the City of Oberlin for the sole purpose of depositing the revenue resulting from the sale of 'green power' attributes, the monies so deposited to be utilised to fund a Sustainable Reserve Program (SRP)." This ordinance established guidelines for the program. Its purpose and goals were also articulated: "provide funding opportunities for community-based, utility-related, environmentally-friendly initiatives demonstrating energy efficiency, energy conservation, green-house gas reductions and/or development of green power generation resources."

The SRP was a grants program administered by the city's electric director. He held the SRP monies tightly and failed to actively promote their use, despite requests from the PUC to do so. I believed that the director saw his job as selling electricity, not promoting sustainability. This belief was confirmed when he passionately stated at a PUC meeting that his role as director was merely to sell electrons, not to foster sustainability.

In April 2015, the PUC asked the city law director to clarify the legal uses of REC money because the electric director believed it belonged to ratepayers. In a written reply, the law director stated: "Based upon the authority of the decision of *Union Ice*, I conclude that the city has no legal obligation to electric customers in the disposition of the REC revenue, and council may, if it wishes, adopt the recommendation of the PUC." That is, the REC revenue belonged to the city, not to OMLPS or the ratepayers.

In May 2015, the electric director recommended to city council that all REC revenue be used to lower electric rates. The PUC majority, on the other hand, recommended that the money be invested in energy-efficiency projects. In response to these divergent recommendations, the city council requested that the electric director and the PUC provide city council with a comprehensive policy for the use of current and future REC revenue.

In May 2015, when these recommendations were made, the PUC was unaware of the city council's 2007 SRP ordinance—despite the fact that the current electric director had written it. In July 2015, a city council member brought it to my attention, and I informed the PUC. The majority of the PUC believed the disposition of REC revenue was clarified by this language: "The Sustainable Reserve Program Account [was established] for the sole purpose of depositing the revenue resulting from the sale of 'green power' attributes, the monies so deposited to be utilised to fund a Sustainable Reserve Program."

The PUC asked the law director if the ordinance was in effect—it was—and if all RECs revenue needed to be deposited in the SRP account. The law director responded that all RECs revenue *could* be put in the SRP account but that there was no requirement for it to be put there. His interpretation of "depositing the revenue resulting from the sale of 'green power'" did not equate to "*all* RECs revenue." Nevertheless, that is exactly what the words in the SRP ordinance state. However, the law director's conclusion implied that the SRP ordinance applied exclusively to the RECs revenue from the college. This is contrary for ordinances in general; ordinances are usually inclusive, not exclusive, as was the ordinance for net metering that was written not only for the first resident with a PV system in Oberlin, but for all who sought to have solar electricity. In my judgement, politics and beliefs played a part in the law director's interpretation, which affirmed those held also by the city manager and electric director.

At the September 2015 meeting of city council, the electric director proposed that 85 percent of the REC revenue be used to lower electric rates and 15 percent deposited in the SRP account. The PUC majority then recommended that all the funds be invested. The city council leadership tried four times to vote on the director's recommendations, but citizen pressure prevented action. Finally, in June 2016, with an unscheduled vote, the city council adopted the director's recommendations by a 4 to 3 vote. The coal dust trio, plus one, still ruled.

Communities for Safe and Sustainable Energy, formed earlier by Oberlin residents, put on the November 7, 2017 ballot, an initiative revising the ordinance to clarify that all net RECs proceeds be deposited in the SRP account. It was approved by 70 percent. Two members of the coal dust trio were term-limited, and the city elected one new council member who had never been on council. The council now had a four-person majority who understood the need to address climate change seriously.

As the PUC majority iterated at many meetings, investing RECs dollars well has the capacity to generate more savings than the value of the REC dollars themselves, especially if matching funds are secured from local foundations and other sources. Such investments are the right thing for the city; they can be used not only to address inequalities in Oberlin—such as to weatherise and in other ways reduce energy use in houses most in need—but also to implement the city's Climate Action Plan, which will accrue untold benefits to Oberlin and beyond.

*****

In fall 2017, I received an unexpected request by two friends from my graduate school days. Would I come to New Hampshire to speak at Star Island's 2018 Natural History Week? My response was like that of Brer Rabbit when threatened by Brer Fox: "Please, do not throw me in the brier-patch." I enthusiastically said yes!

These friends had followed our building and the performance of Trail Magic and believed it to be a perfect fit for Star Island's commitment to becoming a sustainable conference centre. "Building Durable Communities: Lessons for a Liveable Planet" was the theme for the week. Durable communities combine two concepts: durable—stable and continuing to exist—and communities—groups of organisms, such as animals (including humans), microbes, and plants living together in the same environment. These concepts frame the fundamental challenge before humanity. Can we create and sustain durable communities in which humans can fulfil their highest ethical aspirations as articulated by humanity's world religions? After all, these religions foster the most admirable of a culture's accomplishments: ethical standards of behaviour, assistance and comfort to the sick and distressed, rites of passage during life and in death, exemplary arts and music, a purposeful existence, and an atmosphere in which altruistic behaviour thrives: empathy, concern, and help for others.

This invitation provided me the opportunity to employ my scientific perspective on the future to sharpen the focus of thoughtful people seeking to foster sustainable lives and communities. I wove together all that I had learned before and in retirement to present in five two-hour morning sessions. The final presentation considered the complexity of shifting from an economically centred society to an ecologically centred one. All participants were given and asked to read *At the Mercy of Nature: Shackleton's Endurance Saga Gives Promise to Our Future*, my fourth book, in preparation for the last session.

The survival of Shackleton's *Endurance* party is the most notable survival epic in history. Shackleton's initial plan was to sail the *Endurance* from South Georgia Island in the southern Atlantic through the Weddell Sea to Vahsel Bay on the north side of Antarctica where a team of six men would trek across Antarctica via the South Pole. In Shackleton's words, "There now remains the largest and most striking of all journeys—the crossing of the continent." However, 60 miles from Vahsel Bay *Endurance* was beset in Weddell Sea ice and drifted with the ice for nine months until it was crushed and sank. Living on the ice floes for six months, then the *Endurance* party reached the Bransfield Strait that bridges the gap between the Weddell Sea and the Drake Passage. Taking to their three lifeboats for six days, they crossed Bransfield Strait only to be marooned on Elephant Island, an island no one visits. With no other option for rescue, Shackleton and five others sailed their largest life boat, the *James Caird*—a mere 22.5 feet long, six feet wide at the beam, and three-foot seven inches deep—800 miles in the Drake Passage, which has the worse seas in the world, to hit a 25-mile wide target, South Georgia Island. Forced for lack of water to land on the uninhabited side of the island, Shackleton and two others crossed in a forced march the unmapped island to Stromness whaling station, where they received help for rescuing the 22 men still stranded on Elephant Island.

Many people have asked how the 28 men survived when they should have died and become a footnote in the history of Antarctic exploration. The reasons—leadership, discipline, optimism, skills, talents, teamwork, and good luck. That said, there are two critical but contradictory messages for us to understand and act upon if we are to emerge from the 21st century without collapsing our civilisation. First, Shackleton's crew survived because he and other leaders used the best science available when making important decisions. Second, Shackleton absolutely believed that he would bring his men home alive, and they absolutely believed he would do so, despite the lack of any scientifically based evidence to support their beliefs. This contradiction indicates that in dire straits, not only is the best science absolutely essential, but appropriate beliefs are as well.

Shackleton's leadership exhibited six other elements that we would be wise to emulate as we strive to shift from an economically based pattern of living to an ecologically based one. First, equitable pay. His pay ratio was one (crew) to three (scientists). Today the most equitable societies deliver the highest quality of life, especially in terms of education, health care, and safety. Second, fairness in all matters. 'Fairness' did not always mean equal but fair and forthright, as, for example, in food distribution. The food brought to each tent was divided carefully into equal portions, one for each man. One tent mate closed his eyes and turned his head away and called names randomly as the cook for the day pointed at a bowl and said, "Whose?" Portions may not be equal, but allocation was fair. Third, consistent optimism, Shackleton's public face never wavered. He and his crew were going to beat the odds, and they stayed optimistic. Fourth, only take risks when necessary. Demonstrate carefulness, thoughtfulness, and risk-averse behaviour, yet, in tight situations, big risks may be the only option. Fifth, be goal-oriented and have a purpose and meaning; in his case, to cross Antarctica, and when that failed, survive and return home. Sixth, the opportunity for leadership is its own reward—shun personal reward and champion equitable treatment.

*****

Human behavioural history is also relevant. Human groups have a strong propensity for hierarchy and rules, yet, for 95 percent of our existence, we were hunter-gatherers who were largely egalitarian and spiritual. Spirituality and science have in common a propensity to ask questions, including the three most profound: Who are we? Where did we come from? What is our purpose? Myriad answers have been given since we acquired language in the past 100 thousand years. Over the last 10 to 20 thousand years, most of our ancestors adopted an agricultural lifestyle and settled down to occupy a particular place. Religions of various stripes and colours emerged, each offering different answers to the above three questions. Each religion claimed its answers to be true. We have a conundrum! Each of these three questions can have but one correct answer, yet each religion believes its answers to be the correct ones.

Over the past few millennia, the natural sciences emerged with the capacity to provide humanity with the most powerful and elegant way of asking questions and authenticating answers. However, an answer is always conditional and subject to new knowledge. Science has provided well-grounded answers to the first two questions above. Who are we? *Homo sapiens* is one of 5,000 mammalian species extant on Earth in a solar system located in the Milky Way galaxy. Where did we come from? *Homo sapiens* evolved by the process of Darwinian natural selection over the past 3.5 billion years from the first cells that likely evolved by the same process on Earth.

The third question—What is our purpose?—is perhaps more value-dependent than the first two. Being the only species on Earth capable of knowing it exists and of asking questions, humans may believe that their meaning or purpose differs from that of other species. That is, we are unique among all species in some way. Our complex, nuanced language is what likely makes us unique and what has enabled humanity to become the most powerful evolutionary force on the planet. Our activities have unleashed a rapid warming of the Earth as well as the sixth mass extinction of multicellular life in the last 500 million years. Many have written about this role we have taken. Perhaps this is our purpose. If it is not, time is running out for us to do otherwise.

Coda

We are the only species that employs words to tell stories. Although no one knows how long we have been telling stories, the practice began haltingly, deep in our past. Being toolmaking, social animals, we have been communicating with sounds, gestures, and facial expressions for millions of years. Some two million years ago humans began knapping pieces of flint to create spear and arrow heads, an art that had to be facilitated by one person to another. Human capacity for abstract thinking and symbolism likely evolved tens of thousands of years ago as affirmed by art discoveries: red lines drawn on a stone in a South African cave some 75,000 years ago, Sulawesi Cave Art of hand stencils and a carved figure, Venus of Hohle Fels—each about 40,000 years ago.

How reliable is our memory of some past event? Almost two decades ago, when I interviewed Wes Jackson who coined the term 'sustainable agriculture', I asked him what led him to create a particular type of agriculture that did not exist—one for fruit and seed that employed polyculture, was herbaceous and perennial, and mimicked a tall-grass-prairie ecosystem? When I told him that the story, he told me differed from that he had told others, he responded, "Carl, it is different every time I tell it." Jackson had affirmed that almost all memories change over time. When the brain calls up a complex memory, some detail is missing that the brain just fills in with something that fits, thereby making the story mostly accurate, but modified. This happens similarly in a circle of people when a message is passed around ending up different from what was initially said.

Fallibility of memory is perhaps the paramount lesson learned in writing *Beauty Won Me Over: A Scientist's Life*. When I started working in Sussex's lab in 1973, I applied for funds from Yale's American Cancer Society grant. For decades I would have sworn that I received $3,000. When I went to the original paperwork, I discovered that it was $1,500! Yes, my memory was accurate but imprecise—I did apply for money from Yale's American Cancer Society grant, but the award was not $3,000 despite my confidence in that part of the memory. Essentially, all the stories in the foregoing chapters were either not remembered at all or imprecisely. I had saved in 27 Xerox boxes almost all documents from my scientific life: correspondence; grant applications written and reviewed; articles written and reviewed; meetings, symposia, and events organised; materials related to books written, sabbaticals taken, newsletters I wrote and edited, and editorships; and research notebooks. These materials undergird what I have written except for times prior to graduate school, for which I employed documents that I had saved: personal correspondence, school report cards, family letters and records.

I did, however, write of two events for which I have no written confirmation. Each lasted less than a minute, but they were seared into memory—I believe the visual images are as sharp today as they were when burned into memory: shooting down that Connecticut hill on US route 6 headed for West Haven on July 4, 1963, and the moment on Anderson's bridge when I saw the closing silhouette of the Warden off our starboard bow shall be with me until I die for they were the two most exigent moments of my life. Why I responded, as I did will never be known, but because I did, I lived.

Becoming a scientist, in contrast, was far more convoluted and happenstance. My nature and nurture conspired over decades to create the person I am, as they do for everyone. Now that I have told myriad stories that I believe well reflect my life story, perhaps I can tease out the meandering flow down life's river that placed me in the sea of science in general and the gulf of plant development in particular.

I grew up in an established neighbourhood, home of a handful of families with children on the gentle hills above the eastern side of the Tenkill that flowed through the centre of Demarest. We ice skated on the dammed portion in winter and fished up and down the stream on the first day of trout season hoping to catch one recently stocked. I wandered in undeveloped areas of Demarest, alone or with other youths, free of supervision, with hours of time to ponder and explore. Davy's pond a mile east from home with its no longer used icehouse was a frequent destination for catching bull frogs, pollywogs, goldfish, painted turtles, and below the dam the occasional snapping turtle capable of breaking a broomstick with its bite.

In a next-door vacant lot, I had a garden beginning when I was in first or second grade. I grew 'giant' gourds almost too heavy for me to carry that I proudly took to school to show-off along with other plants that could be eaten. Dad had the real vegetable garden in our backyard, a WWII victory garden. When a house was built on the vacant lot with my garden, I was ten and convinced a neighbour who farmed several acres mostly for recreation to let me garden on his land. He agreed to plough a plot for me in return for some of the harvest—I became a sharecropper. Although one never knows, it is likely that these experiences with animals and plants elicited an interest in biology as similar experiences did for many others who also became natural scientists and conservationists.

Going to summer camp for two months when I was 14 as a camper and in subsequent summers as a counsellor in charge of a cabin of boys and instructing activities, I learned how to get along with people in general and take on responsibilities. I was always matter of fact; even as a boy I made some notably serious decisions—such as not to smoke, drink or cheat. As a teenager, I naturally became interested in girls but was cautious about intimate involvement. I dated my share of young women in high school, but Rene left a lasting memory from a party in the classic 1950s basement recreation room with dim lights, soft music, and comfortable furniture. Her hormones were running in overdrive as was her interest in sensual contact. I shied away, fearing the responsibility of fatherhood before its time. Rene was pregnant in her first year at college.

Dad and Mom lived the ethical values they bequeathed me—honesty and truthfulness, do your best, respect elders and authority, honour commitments, and be a responsible citizen. Being Mildred Mosier's boy at the beginning of my teenage years complemented these values, taught me in depth the doing of numerous tasks a homeowner has indoors and outdoors, and frugality. I cannot overestimate the import my parents and Mosier had by instilling not only these behaviours and ethical standards but also their acceptance and respect of me—I knew who I was.

I lack any real musical, artistic, or athletic talents and have an unexceptional intellect. What I do have is mental toughness coupled with the capacity to focus, stay on task, and push limits. I am a high-energy person who likes to talk about things of substance, gravitates to tasks that have obvious goals to achieve, and has a willingness to take on leadership roles. My self-image as a boy was that of a skinny kid standing on the side of the swimming pool waiting to be picked last. Nevertheless, I was, by and large, successful in school, at summer camp, in explorer scouts and Hi-Y. Then, running and Don Gehrmann's coaching enabled me to emerge as a championship athlete, thereby radically transforming my self-image to that of a person who through persistent hard work was capable of accomplishments that others did not imagine possible. Gilbert and Scott assessed that I had no potential as a scientist. Berry wrote in a recommendation letter that I continually had scientific accomplishments he and others thought not possible.

Sussex, like Gehrmann, was uniquely important for my scientific success. His warm welcome into an aggregation of exceptional scientists-in-the-making and his mentorship set me up to succeed—he did not prejudge, rather waited to see what a person did. I had graduated to being judged by the metric employed for assessing scientists: quality of what was published. My Sussex postdoctoral led to three publications: a journal article from a symposium and two articles in *Nature*, one of the most prestigious general science journals in the world. Sussex's aspirational letter of recommendation to Rensselaer's search committee secured a job offer I was pleased to take when Missouri's position vanished with a hiring freeze.

At heart, I was a developmental biologist enchanted by Carlson to pursue somatic cell genetics when a chance observation while creating haploid tobacco plants allowed me because of my training in animal development to recognise that plants counted nodes. How tobacco plants counted nodes prior to forming their terminal flower became a twenty-five year quest during which I applied the animal concept of determination to plant development and the regulation of flowering.

When I joined the Sussex lab, he was one of the few plant scientists who thought developmentally. In animal biology, development was a major field while botany only had physiology, a way of thinking that, by and large, lacked developmental perspective. I had joined the vanguard of Sussex students—

graduate, post-doctoral, and visiting faculty numbering almost 100; a dozen or so came to dominate plant developmental biology in the 1980s and 1990s.

At first, I could not get node-counting research proposals funded because the plant physiologists who reviewed my submissions did not understand them. The same was true of the botanists who reviewed my early papers as illustrated by the initial rejection of "Determination of growth pattern in axillary buds of *Nicotiana tabacum* L." by *Developmental Biology*. When the manuscript was finally accepted, I was bold and sent a reprint to Anton Lang. His positive response led me to send the next developmentally based flowering manuscript to Lang for publication in *Planta*. Its publication there gave me three developmentally focused node-counting papers in prestigious journals—*Nature* (1976), *Developmental Biology* (1978), and *Planta* (1980). I secured NSF funding in 1982 based on these publications and the presentation in January 1979 of my most well received research seminar to the biology department at the University of California-Irvine on the invitation of Bryan, who would soon oversee NSF's developmental biology panel. NSF and USDA would fund my flowering research for more than a decade.

My developmental perspective gained international recognition when I was invited to write review chapters for two prestigious books published in 1984: *Shoot Meristem Development* in *Positional Controls in Plant Development* (Cambridge University Press) and *Induction, Competence and Determination* in *Plant Development in Pattern Formation: A Primer in Developmental Biology* (Macmillan). A decade after I had observed that tobacco plants count nodes, leaders in the international botanical community had accepted my developmental perspective. I had by Wilson's definition become a real scientist: I discovered the fundamental developmental processes by which a tobacco plant counts nodes and becomes determined to flower.

My experience with Lloyd Evans and Rod King demonstrated how difficult it is to broaden one's perspective. We spent tens of hours during my first months in Australia discussing the capacity of a developmental perspective to elucidate aspects of the flowering process not revealed by a physiological perspective alone. When I initiated writing the paper on my project of culturing *Lolium* apices, we agreed on publishing in *Developmental Biology* and highlighting when apices were determined to flower. When I returned to Troy, they reverted to being physiologists. It took a year and many faxes to reinstate a developmental perspective into their physiological viewpoint and publish in *Planta*.

My early node counting studies led to an invitation to attend the first few Plant Cell and Tissue Culture Gordon Conferences at which I met a brilliant young botanist, Michael Christianson, who also had a developmental perspective. We, with Dan Walker, initiated the *Plant Developmental Biology Newsletter* in 1986. *PDBN* had two or three issues each year for eight years that were funded by generous donations from many of the 300 who received the newsletter. The publication helped create and brought together the emergent plant developmental biology community. Innovative ideas were born and brought to fruition—a list of volunteer pre-submission reviewers who significantly improved submitted manuscripts, promotion of a simple three category ranking of research proposals, providing commentaries on classic and new publications and books, and stimulating various societies to include plant developmental symposia in their annual meetings.

My experimental scientist worldview permeated my life. It caused me to experiment with approaches to assessing what students learned in my classes that culminated in a scientifically based assessment that three colleagues and I made of a traditional, instructor-centred, lecture-homework-exam format verses student-centred, Web-based, interactive pedagogy that helped establish just how poor the traditional lecturing format was in comparison.

In community service my science grounding made me chafe at economists' assumption that the environment was part of the economy. Gowdy was a superb facilitator of my economic education that fostered our writing of *Paradise for Sale: A Parable of Nature*. E.O. Wilson had earlier become a person for whom I had immense admiration and respect. Because he helped us secure a publisher for *Paradise for Sale*, he became another eminent scientist whom I was privileged to know professionally and personally.

My last face-to-face interaction with Wilson occurred after he gave a talk to about a thousand at Williams College in Williamstown, MA. Following the talk, he sat at a table to sign books and meet people. I hustled to be close to the head of the line. We talked for several minutes and he signed first editions of *Sociobiology* and *Naturalist*. I then gave him a jar of our blackcap jelly that he had enjoyed in the past, picked up the signed books and said goodbye. As I approached the exit door, I reached in my coat pocket for my gloves and found the jar of blackcap jelly that I had taken along with the signed books. I went back to the table and slowly eased up to Wilson's side and quietly put the jar on the table. He caught me in his peripheral vision and turned. Looking at me with a 'got ya' expression on his face said, "There's the Indian giver." When he graciously provided a book blurb in 2014 for *At the Mercy of Nature*, I sent him another jar that was shared with those in his laboratory.

We retired to Oberlin, OH, in 2008 and had built Trail Magic, a positive energy home that runs on sunshine, producing and acquiring more energy than it uses as our contribution to David Orr's effort to make Oberlin a durable community. Don Watson, my colleague from Rensselaer who did the schematic design for Trail Magic, considers it unique not because it runs on sunshine—many homes do—but because its performance has been extensively documented with measurements and experiments.

We designed Trail Magic's landscape to retain on site the precipitation from a 24-hour, 3-inch rainfall; to enhance biodiversity; to sequester carbon; for food, entertainment, exercise, and ascetics. I am most pleased with the tall grass prairie's performance for enhancing biodiversity—insects, birds, small mammals—and water retention. The swale down the eastern side of the property is tall grass prairie from the north side of the house to the meadow. With a 4-inch rain, no water reaches the meadow. In time the roots of the tall grasses—big blue stem, switch grass, and Indian grass—will achieve a depth of six feet and make rich soil similar to that found in the mid-western US. When Iowa sod was broken over 150 years ago, the tall-grass-produced soil was 12 feet deep and among the best farming land in the world.

Simply stated, Mary and I love living in our Oberlin home and delight every day as we eat meals in our living-dining room and look out onto prairie, pond, meadow, woods, and Plum Creek flood plain. I am provided with exercise everyday—climbing stairs to my study, on the sunless cold days hauling firewood from under the deck up the ground floor stairs to the wood bin in the living-dining room, downing trees that are dying or dead and then cutting and splitting a cord into woodstove size each year, using my 23 horsepower Kubota tractor to mow fields and brush-hog the wilder patches, having a garden, and just walking the property or sitting somewhere to see and listen to nature in all her seasons. Soon after we moved in a person coming to one of our open houses asked how I was feeling about Trail Magic—then and now, like I had died and gone to heaven.

Over the fifty years that I practiced science the world in general, and the US in particular, have drifted away from employing established science as the touchstone for making effective policy. When belief-based policy replaces science based policy, as it did in the Soviet Union from the early 1930s to the mid-1960s when genetics was abandoned for Lysenko's pseudoscience, millions of people die and others suffer unnecessarily. During the Presidencies of George W. Bush and Donald J. Trump the US has selectively abandoned science for ideology to the detriment of effective policy. The most egregious for humanity's future have been overt actions backed by billions of dollars to deny that burning fossil fuels is the primary cause of the ever-increasing atmospheric heat-trapping gases that have brought climate chaos upon civilisation, whose resilience will be severely tested.

On the brighter side I have been privileged to be a scientist and to interact professionally and personally with innumerable accomplished pliers of the trade. My foremost mentors—Berry, Lang, Sussex—have passed on. Yet each is deeply imprinted into my very being and finds expression in all I do. Did I choose to be a scientist? Probably not, although at this point it is of little concern. My life is realised.

CPSIA information can be obtained
at www.ICGtesting.com
Printed in the USA
LVHW081527200621
690707LV00002B/201

9 781528 986038